# The Physics of Advanced Optical Materials: Tellurite Glasses

Raouf El-Mallawany

# The Physics of Advanced Optical Materials: Tellurite Glasses

Springer

Raouf El-Mallawany
Physics Department, Faculty of Science
Menoufia University
Shebeen ElKomm, Egypt

ISBN 978-3-031-45244-4 ISBN 978-3-031-45245-1 (eBook)
https://doi.org/10.1007/978-3-031-45245-1



This Springer imprint is published by the registered company Springer Nature Switzerland AG
The registered company address is: Gewerbestrasse 11, 6330 Cham, Switzerland

Paper in this product is recyclable.

*Dedicated to my family*

# Preface

Why it is important to study tellurite glasses?

Because tellurite glasses are a new kind of glass with unique physical properties and sophisticated applications (as explained in detail at the end of the book).

Why doping tellurite glasses by rare earth (RE) or rare-earth nanoparticles (RENPs)?

To get new optical characteristics of tellurite glasses.

N.B.: a detailed answer will be at the end of Chap. 4.

The motivation of this book is to focus on the optical properties of new tellurite glasses with the impact of rare-earth nanoparticles (RENPs) or transition metal nanoparticles (TMNPs). The optical properties will be discussed in both visible and UV spectra for new tellurite glasses as shown in the figure below. Each chapter will cover introductory, experimental techniques, quantitative, and some crucial data. Also, this book collected recent data according to the research engines and appeals to a broad audience of academic researchers and industry professionals.

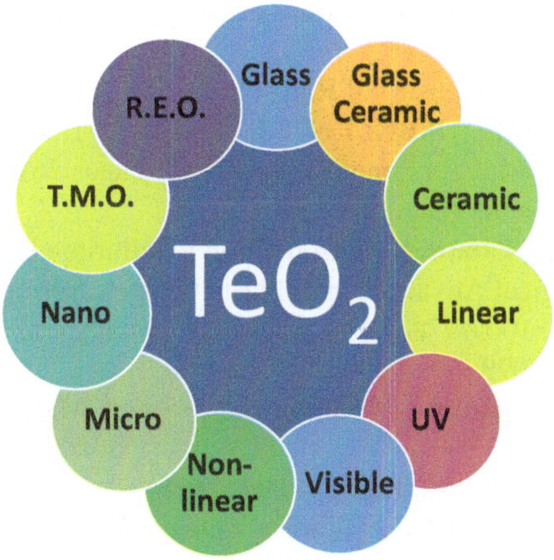

Chapter 1 gives an introduction to the new tellurite glasses, glass ceramics, and transparent ceramics for different purposes. Moreover, it gives a review of the recent application and challenges of the three kinds of tellurite glasses. Also, it shows the optical properties of tellurite glasses in the visible region and divided this task into sub-directions: some experimental data of linear refractive index and theoretical analysis of the linear refractive indices. Also, the universal power law of terahertz optical properties has been collected and discussed.

Chapter 2 contains the nonlinear optical properties of tellurite glasses. While, Chap. 3 focuses on the interaction of the UV spectra with tellurite glasses by the absorption spectra fitting (ASF) method and differential absorption fitting (DASF). Moreover, it contains measurements of UV spectra, theoretical background, experimental data, optical band gap, the impact of RENPs, transition metal-doped tellurite glass ceramics, and quantum dot-embedded tellurite glass. Then, Chap. 4 presents some applications of tellurite glasses, glass ceramics, and ceramics for different purposes.

Besides the previous, the present book will complete the previous series of books on tellurite glasses and the special issue *Journal of Non Crystalline Solids*:

1. RAH El-Mallawany, 'Tellurite Glasses Handbook, Physical Properties and Data', First Edition: https://doi.org/10.1201/9781420042085, eBook ISBN: 9780429129124.
   Published by CRC-Press, USA, 568 Pages (2002).
2. RAH El-Mallawany, 'Tellurite Glasses Handbook, Physical Properties and Data' Second Edition: Published by CRC-Press, USA, 502 Pages (2011).
   ISBN: 978-143984984-2, 978-143984983-5.
3. The **5** video educational lectures for the International Materials Institute for New Functionality in Glass at Lehigh University, Bethlehem, PA, USA (2005) https://www.lehigh.edu/imi/teched/LecBasic/IndvLecturesBasic.
4. NS Hussain, **R. El-Mallawany**, G Hungerford, MJM Gomes, SK Mendiratta, **Chapter 8**: Optical Properties of Rare-Earth Ions Doped Lithium Boro Tellurite Glasses,
   **Book**: Physics and Chemistry of Rare-Earth Ions Doped Glasses,
   ISBN: 0-87849-4881-2, ISBN: 13978-0-87849-481-1,
   ISBN: 1422-3597 Mat.Sci.Found.,Vols. 46–47, Pages 261–286.
   TTP publishers, Switzerland (2008).
5. **R. El-Mallawany**, **Chapter 1**-Introduction to tellurite glasses,
   **Book**: Technological Advances in Tellurite Glasses, Springer 1–13 (2017), https://doi.org/10.1007/978-3-319-53038-3, Library of Congress Control Number: 2017935099.
6. **R. El-Mallawany**, **Editor**, Tellurite Glass Smart Materials, Applications in Optics and Beyond.
   ISBN: 978-3-319-76568-6, Pages 1–297 (2018).
7. **R. El-Mallawany**, **Chapter 1**: Tellurite Glass Smart Materials: Applications in optics and beyond

**Book**: Tellurite Glass Smart Materials, Springer, 1–16 (2018), (Springer) (http://www.springer.com/us/book/9783319765679) ISBN: 978-3-319-76568-6, Pages 297 (2018)

8. **R. El-Mallawany**, **Chapter 2**: Radiation Shielding Properties of Tellurite Glasses,
   **Book**: Tellurite Glass Smart Materials, Springer 17–27 (2018), (Springer) (http://www.springer.com/us/book/9783319765679) ISBN: 978-3-319-76568-6, Pages 297 (2018)
9. **R. El-Mallawany**, **Preface**, Journal of Non Crystalline Solids Special Issue: "Tellurite Glasses, properties, Structure and Applications", 528, 119784 (2020). 10.1016/j.jnoncrysol.2019.119784.

Shebeen ElKomm, Egypt Raouf El-Mallawany

# Contents

# Chapter 1
# Linear Refractive Index of Some Tellurite Glasses

The present chapter deals with the refractive index of tellurite glasses. It is divided into two sub-directions: experimental data of linear refractive index and theoretical analysis of the linear refractive indices. Also, refractive index in terahertz (THz) has been reported. Three significant material classes for a wide range of applications are tellurite glasses, tellurite glass ceramics, and tellurite ceramics. The theoretical analysis of linear refractive indices and experimental data related to them are covered in this chapter. Theoretical interpretation for linear refractive index, Sellmeier coefficients, the effect of nano-rare-earth particles on linear refractive index, and refractive index in terahertz (THz) have been collected and discussed. Moreover, the thermo-optic coefficient has been collected.

In the 1950s, tellurite glasses were established [1]. After that, there were few publications throughout the 1960s, 1970s, and early 1980s [2–4]. High-purity tellurite glass was made possible in 1984 [4]. There have been numerous reports of tellurite glasses treated with R.E.O. or T.M.O. [5–11]. For many applications, tellurite glass ceramics almost emphasize the doping of RE ions or RENPs. There have been hundreds of scholarly articles produced annually since the 1990s. New tellurite glass systems doped with rare-earth (RE) ions and with rare-earth nanoparticles (RENPs) tellurite glasses have been achieved [12–17].

More than that, spectroscopic analyses of $TeO_2$–$WO_3$ glasses doped with $Dy^{3+}$ for solid-state lighting applications have attracted numerous researchers [18–58] to tellurite glasses in 2023. Creating and characterizing $Eu^{3+}$-doped tellurite glasses for laser materials has been reported [54].



R. El-Mallawany, *The Physics of Advanced Optical Materials: Tellurite Glasses*,
https://doi.org/10.1007/978-3-031-45245-1_1

## 1.1 Experimental Measurement and Some Data of Linear Refractive Index

The refractive index has been measured using a traveling microscope and a spectrophotometer (Novaspec II Visible Spectrophotometer) with a manually adjustable light beam wavelength as shown in Fig. 1.1. Vernier readings of 0.01 mm were employed, and two slow motion knobs were included with the traveling microscope.

Refractive index ($n_\lambda$) varies with wavelength ($\lambda$) according to the next formula:

$$n_\lambda = A + \frac{B}{\lambda^2} + \frac{C}{\lambda^4} + \cdots \tag{1.1}$$

*A*, *B*, and *C* are constants. From this relationship, the refractive index increases with decreasing wavelength. Also, Abbe number has been defined as:

$$\nu_d = \frac{n_D - 1}{n_F - n_C} \tag{1.2}$$

where, $n_D$, $n_F$, and $n_C$ = refractive indices for the material at the wavelength of Fraunhover *c*, *d*, and *F* spectral lines (656.3, 587.56, and 486.1 nm, respectively). Generally, Abbe number varies oppositely with the refractive index in optical glass.

Molar refractivity $R_M$ has been calculated by using "Lorentz–Lorenz" equation

$$\left\{\frac{n2-1}{n2+2}\right\}\left\{\frac{M}{P}\right\} = \mathrm{RM} = 4\pi \ \mathrm{NA} \ \alpha_m/3 \tag{1.3}$$

($M$) = molecular weight of the materials, ($p$) = density, ($\alpha_m$) = molar polarizability and ($M/p$) = molar volume $V_m$, and $N_A$ = Avogadro's number. The Lorentz–Lorenz equation allows calculating the electronic polarizability of oxide ions, $\alpha_{O_2^-}$ ($n$), in oxide materials as indicated in Eq. (1.4):

$$\alpha_{O_2^-}(n) = \left[\frac{R_m}{2.52} - \Sigma\alpha_i\right]\left(N_{O_2^-}\right)^{-1} \tag{1.4}$$

where molar refraction, $R_m$, $\Sigma\alpha_i$ = molar cation polarizability, and $N_{O_2^-}$ denotes the number of oxide ions in the chemical formula. Also, the optical basicity of the oxide medium, $\Lambda$, can be calculated by Eq. (1.5),

$$\Lambda = 1.67\left(1 - \frac{1}{\alpha_{O_2^-}}\right) \tag{1.5}$$

Optical basicity $\Lambda_{\mathrm{th}}$ of multicomponent glass by Duffy [59] is shown in Eq. (1.6),

$$\Lambda_{\mathrm{th}} = X_1\Lambda_1 + X_2\Lambda_2 + \cdots X_n\Lambda_n \tag{1.6}$$

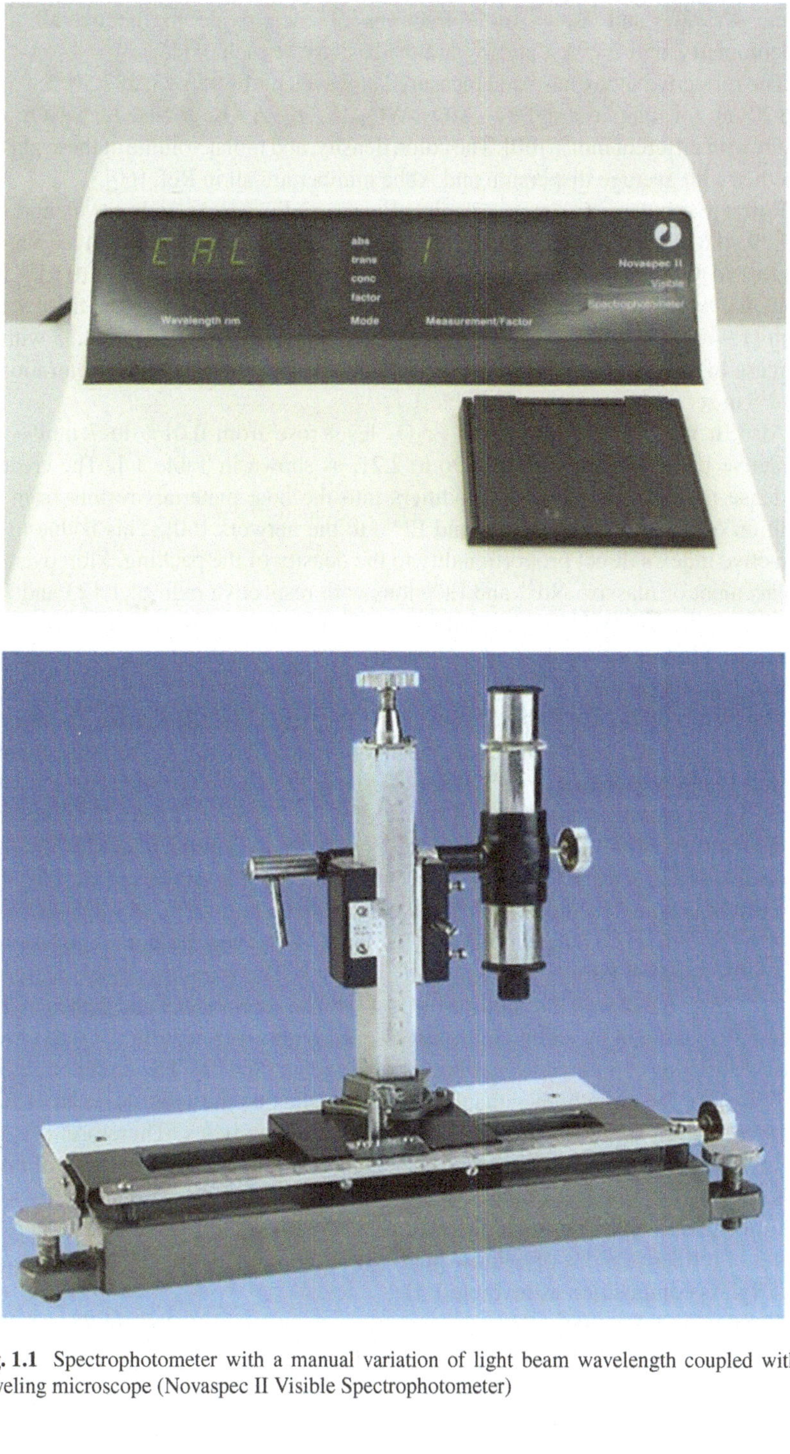


**Fig. 1.1** Spectrophotometer with a manual variation of light beam wavelength coupled with a traveling microscope (Novaspec II Visible Spectrophotometer)

where $X_1$, $X_2$, and $X_n$ = mole fractions of each oxide to the overall glass stoichiometry, and $\Lambda_1$, $\Lambda_2$, and $\Lambda_n$ are the basicity of each oxide.

The refractive index has been measured at wavelengths 486.13, 587.56, 589, and 656.27 nm for the glasses: $TeO_2$–$TiO_2$–$WO_3$–$A_nO_m$. $A_nO_m$ = $Nb_2O_5$, $Nd_2O_3$, and $Er_2O_3$ with different mol% [60]. The color, density, and molar volume of these glasses together with average dispersion and Abbe number are all in Ref. [60].

Refractive indices ($n$) were calculated using sodium light with wavelengths of 589.29, 486.13, 587.56, 589, and 656.27 nm for the three glass series (A) for $Nb_2O_5$, (B) for $Nd_2O_3$, and (C) for $Er_2O_3$. With increasing modifier levels from (0.01% to 7 mol%), it was discovered that values of $n$ (589.29 nm) increased in the three series from (1.94 to 2.22). The linear refractive index increased from 2.10 to 2.15 with an increase in $Nb_2O_5$ content from 0.01% to 5 mol%, considering the concentration of ($Nb^{5+}$) in series ($A$) as shown in Table 1.1.

Also, it was discovered that as $Er_2O_3$ level rose from 0.01% to 7 mol%, the refractive index increased from 1.96 to 2.21, as shown in Table 1.1. The creation of dense packing of rare-earth modifiers into the host materials results from the addition of rare-earth ions ($Nd^{3+}$ and $Er^{3+}$) to the network [60]. This is due to the refractive index's direct proportionality to the density of the packing. Moreover, the replacement of massive $Nd^{3+}$ and $Er^{3+}$ ions with respective radii of 1.123 and 1.03 with a smaller ion ($W^{6+}$) with a radius of 0.74 [60] increases the refractive index. These are brought on by the high polarization values that these ions possessed as a result of an increase in cation size.

The Abbe numbers ($\nu_d$) were discovered in the range of 18.05–25.16. The average dispersion ($n$F–$n$C) for the three series are collected in Table 1.1. The Abbe number ($d$) was found to decrease for the ($A$) series from (23.69 to 16.02) as the concentration of $Nb_2O_5$ increased from (0.01 to 5 mol%), for the ($B$) series from (25.16 to 16.02 as the concentration of $Nd_2O_3$ increased from (0.01 to 7 mol%), and for the third series from (22.78 to 18.05) as the concentration of $Er_2O_3$ increased. Besides that, the current chapter's conclusion will include a recent study of these values.

Also, the refractive index of ($TeO_2$)–($B_2O_3$)–($Gd_2O_3$) was measured at 632.80 nm [62] with standard error $\pm$ 0.1 nm as in Table 1.1. The refractive index decreased from 2.029 to 1.940 with the addition of $Gd_2O_3$ to the glass system and then increased from 1.940 to 2.058 by adding more $Gd_2O_3$. Bridging oxygen atom (BO) has lower polarizability as compared to non-bridging oxygen atoms [9]. The atomic radius of $Gd^{3+}$ ions (1.79 Å) is higher than the atomic radius of the tellurite atom (1.6 Å) and boron atom (0.98 Å). This improves the refractive index. Therefore, ($TeO_2$)–($B_2O_3$)–($Gd_2O_3$) glasses have sophisticated applications in nonlinear optical materials. Figure 1.2 represents values of the refractive index against density for some tellurite glasses. Glasses with higher density give a higher refractive index. Also, the values of refractive index of tellurite glass with nanoparticles are higher the ordinary glass (glass composition as in Table 1.1).

The molar refractivity $R_M$ values were in the range: 16.268–16.77, 15.591–17.02, and 14.032–16.50 $cm^3/mol$ for these three families of tellurite glasses doped by $Nb_2O_5$, $Nd_2O_3$, and $Er_2O_3$, respectively. The relationship between the linear refractive index ($n$) and the energy gap $E_g$ will discuss later in Chap. 3.

**Table 1.1** Glass composition, density, refractive index, and Abbe number

| Glass composition | Density g/cm$^3$ | Refractive index | | | Abbe number | References |
|---|---|---|---|---|---|---|
| Glass series A: | | $n$ (486 nm) | $n$ (587 nm) | $n$ (656 nm) | | El-Mallawany et al. [60] |
| $80TeO_2$–$5TiO_2$–$(15 - x)$ $WO_3$–$xNb_2O_5$ | 5.455 | 2.32 | 2.28 | 2.24 | 15.55 | |
| $X = 0.01$ and $X = 1.0$ | 5.49 | 2.18 | 2.09 | 2.02 | 6.86 | |
| Glass series B: | | | | | | El-Mallawany et al. [60] |
| $80TeO_2$–$5TiO_2$–$(15 - x)$ $WO_3$–$x$ $Nd_2O_3$ | | | | | | |
| $x = 0.01$ | 5.214 | 2.08 | 2.04 | 1.94 | 7.6 | |
| $X = 1.0$ | 5.379 | 2.08 | 2.03 | 1.99 | 11.3 | |
| $X = 5.0$ | 5.680 | 2.19 | 2.10 | 2.02 | 6.41 | |
| $X = 7.0$ | 5.2 | 2.01 | 1.99 | 1.97 | 21.37 | |
| Glass series C: | | | | | | El-Mallawany et al. [60] |
| $80TeO_2$–$5TiO_2$–$(15 - x)$ $WO_3$–$x$ $Er_2O_3$ | | | | | | |
| $x = 0.01$ | 5.778 | 2.22 | 2.17 | 2.08 | 7.94 | |
| $X = 3.0$ | 5.887 | 1.89 | 1.85 | 1.83 | 15.13 | |
| $X = 5.0$ | 5.972 | 1.76 | 1.75 | 1.68 | 9.95 | |
| $xLi_2O$–$(100 - x)TeO_2$ | 5.113–4.653 | 2.0793–1.9508 | | | | Honma et al. [61] |
| $xNa_2O$–$(100 - x)$ $TeO_2$ | 5.195–4.811 | 2.0920–1.9870 | | | | |

(continued)

**Table 1.1** (continued)

| Glass composition | Density g/cm$^3$ | Refractive index | | | Abbe number | References |
|---|---|---|---|---|---|---|
| {[($TeO_2$)0.70 ($B_2O_3$)0.30]0.70 (ZnO)0.30}1 − *y* ($Er_2O_3$/$Er_2O_3$ NPs) *y*; Ordinary glass | | | | | | Azaln et al. [28] |
| $y = 0.00$ | 3.630 | 1.700 (632.8 nm) | | | | |
| $y = 0.03$ | 3.790 | 1.730 (632.8 nm) | | | | |
| $y = 0.05$ mol% | 3.960 | 1.740 (632.8 nm) | | | | |
| {[($TeO_2$)0.70 ($B_2O_3$)0.30]0.70 (ZnO)0.30}1 − *y* ($Er_2O_3$/$Er_2O_3$ NPs) *y*; Glass with nanoparticles | | | | | | Azlan et al. [28] |
| $y = 0.00$ | 3.630 | 1.700 (632.8 nm) | | | | |
| $y = 0.03$ | 4.822 | 1.855 (632.8 nm) | | | | |
| $y = 0.05$ mol% | 4.881 | 1.924 (632.8 nm) | | | | |
| 90 $TeO_2$–5 ZnO–5 $Y_2O_3$ | 5.50 | 2.141 (543.5 nm) | 2.112 (632.8 nm) | | | Strutynski et al. [23] |
| | | 2.067 (1064 nm) | 2.059 (1550nm) | | | |
| 70 $TeO_2$–25 ZnO–5 $Y_2O_3$ | 5.44 | 2.064 (543.5 nm) | 2.037 (632.8 nm) | | | Strutynski et al. [23] |
| | | 1.998 (1064 nm) | 1.939 (1550 nm) | | | |
| 55 $TeO_2$–40 ZnO–5 $Y_2O_3$ | 5.36 | 1.997 (543.5 nm) | 1.976 (632.8 nm) | | | Strutynski et al. [23] |
| | | 1.943 (1064 nm) | 1.939 (1550 nm) | | | |
| 80 $TeO_2$–10 ZnO–10 $Y_2O_3$ | 5.42 | 2.082 (543.5 nm) | 1.976 (632.8 nm) | | | Strutynski et al. [23] |
| | | 2.006 (1064 nm) | 1.999 (1550 nm) | | | |
| (1 − *x*) [($TeO_2$)70 ($B_2O_3$)30] − *x* ($Gd_2O_3$) and $x = 0.2$ | 3.914 | 2.029 ± 0.01 (632.8 nm) | | | | Azlan et al. [62] |

(continued)

**Table 1.1** (continued)

| Glass composition | Density g/cm$^3$ | Refractive index | | | Abbe number | References |
|---|---|---|---|---|---|---|
| 0.4 | 4.045 | 1.940 ± 0.01 (632.8 nm) | | | | |
| 0.6 | 3959 | 1.951 ± 0.01 (632.8 nm) | | | | |
| 0.8, mol% | 4014 | 1.964 ± 0.01 (632.8 nm) | | | | |
| $Nd^{3+}$ ions doped oxy-fluoro borotellurite glasses with the composition (20 − $x$) $Na_2O$–5 $BaF_2$–5 $CaF_2$–60 $B_2O_3$–10 $TeO_2$–$x$ $Nd_2O_3$, | | | | | | Krishna et al. [63] |
| $x = 0.5$, | 2.925 | 2.291 | | | | |
| 1.0 | 2.950 | 2.280 | | | | |
| 1.5 | 2.975 | 2.271 | | | | |
| 2.0 | 2.999 | 2.261 | | | | |
| 2.5 mol% | 3.024 | 2.254 | | | | |
| $TeO_2$, ZnO, $WO_3$, $Bi_2O_3$, doped by $Nd_2O_3$, $Tm_2O_3$, $Ho_2O_3$ | | | | | | Ding et al. [64] |
| T-Nd | 5.26 | 2.012 | | | | |
| T-Tm | 5.766 | 2.015 | | | | |
| T-Ho | 5.504 | 2.014 | | | | |
| T-NdTm | 5.753 | 2.013 | | | | |
| T-NdTmHo | 5.659 | 2.013 | | | | |
| $60TeO_2$–(30 − $x$)ZnO–$5Bi_2O_3$–$5TiO_2$-$xB_2O_3$ (where $x$ = 0, 10.0) | 5.8791 | 1.875 ± 0.015 | 1.741 ± 0.021 | | | Marzuki et al. [22] |

(continued)

**Table 1.1** (continued)

| Glass composition | Density g/cm$^3$ | Refractive index | | | Abbe number | References |
|---|---|---|---|---|---|---|
| | 5.6463 | | | | | |
| $(1-x)[70TeO_2$-$30B_2O_3]$-$xGd_2O_3$ | | At (632.8) nm | | | | Azlan et al. [62] |
| 0.3 | 3.914 | 2.029 | | | | |
| 0.4 | 4.045 | 2.940 | | | | |
| 0.6 | 3.959 | 1.951 | | | | |
| 0.8 | 4.014 | 1.964 | | | | |
| $\{[(TeO_2)0.70\ (B_2O_3)0.30]0.7\ (ZnO)0.3\}(1-y)$–$y(Sm_2O_3$ (MPs) | MPs | At (632.8) nm | | | | Azlan et al. [25] |
| 0.0 | 3.693 | 1.871 | | | | |
| 0.01 | 3.709 | 1.974 | | | | |
| 0.02 | 3.714 | 1.873 | | | | |
| 0.04 mol% | 3.795 | 2.030 | | | | |
| $\{[(TeO_2)0.70\ (B_2O_3)0.30]0.7\ (ZnO)0.3\}(1-y)$–$y\ (Sm_2O_3$ (MPs) | NPs | At (632.8) nm | | | | Azlan et al. [25] |
| $y = 0.005$ | 3.72 | 2.056 | | | | |
| 0.03 | 3.817 | 2.045 | | | | |
| 0.04 mol% | 3.951 | 2.050 | | | | |

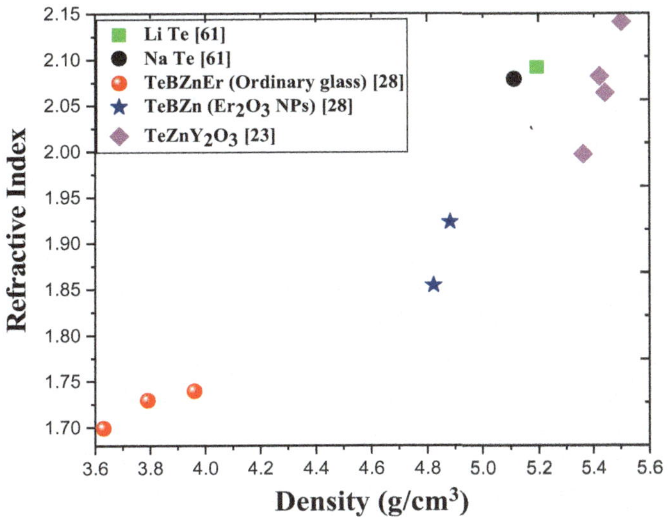


**Fig. 1.2** Refractive index—density for some tellurite glasses and others (Composition as in Table 1.1)

## 1.2 Thermo-Optic Coefficient

Alkali tellurite glasses thermo-optic and electronic polarizability has also been measured in 2010 [61]. Table 1.1 lists the values of refractive index and density for $Li_2O$–$TeO_2$ and $Na_2O$-$TeO_2$ glasses at room temperature. $Li_2O$–$TeO_2$ and $Na_2O$–$TeO_2$ glasses both obtained high refractive indices of $n = 1.95$–$2.07$, electronic polarizabilities 2.26–2.43 with 2.26–2.43, and significant optical basicity of $n = 0.93$–$0.98$ [61]. $Li_2O$–$TeO_2$ and $Na_2O$–$TeO_2$ glasses got high refractive indices of $n = 1.95$–$2.07$, large electronic polarizabilities of oxide ions of $\alpha_{O_2^-} = 2.26$–$2.43$ Å$^3$, and large optical basicity of $\Lambda = 0.93$–$0.98$. The large values of $n$, $\alpha_{O_2^-}$, and $\Lambda$ obtained in $Li_2O$–$TeO_2$ and $Na_2O$–$TeO_2$ glasses suggest that these glasses are basic. The values of $n$, $\rho$, and $\alpha_m$ decrease clearly with increasing either $Li_2O$ or $Na_2O$ content in both glasses. In $Li_2O$–$TeO_2$ glasses, the values of $\alpha_{O_2^-}$ and $\Lambda$ tend to decrease very slightly with increasing $Li_2O$ content. Regarding the $Na_2O$–$TeO_2$ glasses, the values of $\alpha_{O_2^-}$ and $\Lambda$ are almost constant, irrespective of the $Na_2O$ content. The optical basicity of $Li_2O$, $Na_2O$, and $TeO_2$ oxides have been reported to be $\Lambda = 0.87$, 1.15, and 0.99 (or 0.93), respectively [61].

Moreover, the temperature dependence of the refractive index ($n$) of glasses (thermo-optic coefficient) has been performed [61]. By differentiating the Lorentz–Lorenz equation:

$$\frac{dn}{dT} = \frac{(n^2 - 1)(n^2 + 2)}{6n}(\phi - \beta) \tag{1.7}$$

where $\varphi$ is the polarizability coefficient indicating the temperature dependence of the electronic polarizability $\alpha$:

$$\phi = \frac{1}{\alpha}\frac{d\alpha}{dT} \tag{1.8}$$

It can be observed that:

i. If the contribution of volume expansion $\beta$ is large compared to the contribution of polarizability, i.e., $\varphi < \beta$, the sign of ($dn/dT$) would be negative,
ii. In contrast, if $\varphi > \beta$, the sign of ($dn/dT$) would be positive.

For $Li_2O–TeO_2$ and $Na_2O–TeO_2$ glasses, the contribution of polarizability $\varphi$ on the thermo-optic coefficient was calculated, where the values of refractive index at 50 °C were used. The values of $\varphi = 9.02–9.50 \times 10^{-5}\ K^{-1}$ for $Li_2O–TeO_2$ glasses and $\varphi = 8.69–9.45 \times 10^{-5}\ K^{-1}$ for $Na_2O–TeO_2$ glasses were obtained. The contribution of polarizability increases with increasing alkali ($Li_2O$ or $Na_2O$) content. Furthermore, it should be pointed out that the condition of $\varphi > \beta$ was held in all glasses examined in that study [61]. Therefore, $TeO_2$-based glasses got a high potential for devices based on interferometric principles such as optical switches, tunable filters, and fiber–optic sensors that require large thermo-optic coefficients.

Recently in 2022, the glass system $60TeO_2–(30\text{-}x)ZnO–5Bi_2O_3–5TiO_2\text{-}xB_2O_3$ (where $x = 0–10.0$) has been prepared [22]. Refractive indices of all samples were measured at room temperature using Brewster's angle method at a wavelength of 632.9 nm as shown in Table 1.1.

Also, the $TeO_2–ZnO–Y_2O_3$ ternary system's tellurite glasses' physicochemical characteristics and capacity for fiber drawing have been prepared [23] at four different wavelengths (542, 632, 1064, and 1550 nm) on meticulously polished slabs. With an estimated error of 103, the measurements were performed using a homemade prism ($TiO_2$) refractometer. The two-pole Sellmeier equation was then used to fit the refractive index dispersion as a function of wavelength (more on this later); the results are reported in Table 1.6. As $TeO_2$ was elevated from 55 to 90 mol%, the refractive index linearly increased from 1.939 to 2.059 (1.55 m).

Moreover, $(20 - x)\ Na_2O–5\ BaF_2–5\ CaF_2–60\ B_2O_3–10\ TeO_2–x\ Nd_2O_3$ glasses were applied in fiber amplifiers and 1064 nm laser applications [63]. The $TeO_2–ZnO–WO_3–Bi_2O_3$ tellurite glasses with $Nd^{3+}$, $Tm^{3+}$, and $Ho^{3+}$ doping have been synthesized [64]. Also, $TeO_2$, $ZnO$, $WO_3$, $Bi_2O3$, $Nd_2O_3$, $Tm_2O_3$, and $Ho_2O_3$ made up the glass compositions. At a wavelength of 632.8 nm, a Sairon Tech SPA-4000 prism coupler was used to measure the refractive index. The $Nd^{3+}/Tm^{3+}/Ho^{3+}$ tri-doped tellurite glasses demonstrated a potential multifunctional application promise in many domains with doping concentrations of 0.015 mol% $Nd_2O_3$, 0.32 mol% $Tm_2O_3$, and 0.045 mol% $Ho_2O_3$. It is well known that the lone pair of electrons make up NBO. These electrons have looser nuclear charge bonds. Lone pair electrons have a high degree of polarizability, which raises their refractive index. Also, the existence of rare-earth ions may lower the average crosslinking density and then increase the value of refractive index [31].

## 1.3 Theoretical Analysis of Linear Refractive Index

The classic dielectric theory provides several relationships between the refractive index and the density of the solid [65]. At optical frequencies, where the refractive index is measured, the dielectric constant arises almost entirely from electronic polarization. The electronic polarizability ($\alpha$) is expressed by the following expression:

$$\alpha = \frac{e^2}{4\pi^2 m} \sum_i \frac{f_i}{\nu_i^2 - \nu^2} \tag{1.9}$$

where ($e$) = the electronic charge, ($m$) = the electronic mass, ($f$) represents the oscillator strength, ($\nu_1$) = the electronic frequency, and ($\nu$) = the frequency of the incident photons. According to Maxwell's theory, the relationship $n^2$ ($\lambda \Rightarrow \infty$) = $\varepsilon$ holds for infinitely long waves. The Clausius–Mossotti equation is explained in detail [66].

$$\begin{aligned} \frac{N\alpha}{3\varepsilon_o} &= \frac{\varepsilon - 1}{\varepsilon + 2} \\ &= \frac{n^2 - 1}{n^2 + 2} \\ &= N\frac{e^2}{12\varepsilon_o \pi^2 m} \sum_i \frac{f}{\nu_i^2 - \nu^2} \end{aligned} \tag{1.10}$$

where ($N$) is the number of electrons in the solid. As indicated in the above equation., the refractive index ($n$) depends upon the composition of an optical material; the more polarizable the outer electrons, the higher the refractive index.

Consequently, the molar refraction ($R_M$) and index of refraction $n$ depend on the polarizability of the material which in turn is determined, approximately, by the sum of ionic refraction ($R_I$). The value of ionic refraction in crystal compounds depends on:

(1) Electronic polarizability of the ion, which increases as the number of electrons in the ion increases,
(2) Coordination number of the ion,
(3) Polarizability of the first neighbor ions coordinated,
(4) Field intensity ($z/a^2$) where ($z$) is the valence of the ion and ($a$) is the distance of separation.

Values of ionic refraction, ionic polarizability, and coordination number for some cations and some anions are shown in Table 1.2 [65]. If halides are melted with $TeO_2$, evaporation is more or less inevitable, so that ternaries of the volatile component, such as $TeO_2$–$Me_2O$–$MeCl_2$. The optical data listed show that a valuable family of optical glasses has been discovered. To understand how tellurite glasses have a high refractive index, the number of polarizable atoms/unit volume (*N/V*), the electronic polarizability of the ion, and ($\alpha_{ionic}$) must all be considered. The values of ($n$) for $TeO_2$ vary from $n$ (glass) = 2.2 to $n$ (crystal) = 2.37 due to the variation of densities from $\rho$ (glass) = 5.1 (g/cm$^3$) to $\rho$ (Crystal) = 5.99 (g/cm$^3$). The number of atoms per unit volume has decreased from $6.78 \times 10^{22}$ (cm$^{-3}$) for $TeO_2$ crystal to $5.77 \times 10^{22}$ (cm$^{-3}$). By knowing the molar refraction of both $TeO_2$ crystal and glass, the polarizability has been decreased from $7.37 \times 10^{-24}$ (cm$^3$) to $6.96 \times 10^{-24}$ (cm$^3$), i.e., ($\alpha_{cryst}/\alpha_{glass}$) = 1.1. Previous conclusions confirmed that the dielectric constant decreased from 23.44 to 20.1 and the static polarizability decreased also from $2.8 \times 10^{-6}$ (cm$^3$) to $2.2 \times 10^{-6}$ (cm$^3$). The ratio between the density, dielectric constant, and static polarizability of $TeO_2$ crystal to $TeO_2$ glass varies between 1. 18 and 1.27. Thus, tellurite glass has a lower refractive index due to low electronic polarizability and higher specific molar volume. This indicated the very important role of the $TeO_2$ unit in determining the refractive index of binary and ternary tellurite glasses.

The composition of the glass and its density are used to calculate the number of atoms/ unit volume (*N*/*V*) which is dependent on the modifier as in Table 1.2 [66]. The trend of refractive index along with $Nd^{3+}$ content in {[($TeO_2$)0.70 ($B_2O_3$)0.30]0.7 (ZnO)0.3}$1 - y$ ($Nd_2O_3$)$y$ and $y$ increased from 0.005 to 0.05; was an increase of $Nd^{3+}$ from 1.760 to 1.863 [30]. The enhancement of the refractive index can be explained by the role of $Nd^{3+}$ to alter the structural properties of the borotellurite glass system. The large value of optical basicity in the glass series revealed that the glass system is more basic. Oxide ions in the glass system act as a Lewis base which donates electrons to the surrounding cations. The high degree of donating ability of oxide ions influences strongly the acid–base characteristic.

Strong oxide ion donation to the surrounding cations is facilitated by low polarizing cations in the glass system. There is good agreement with optical basicity based on the prior results on oxide ion polarizability. The glass system exhibits a nonlinear trend in optical basicity and concentration. The structural change in the glass system is what is responsible for the nonlinear trend in optical basicity. In addition, it has been observed that the optical basicity values somewhat rise with increasing $Nd^{3+}$ concentrations. By contrasting the value of the optical basicity of various glass system components, this trend can be explained. Table 1.3 [67] lists the values for the metallization criterion of borotellurite glasses that have been doped

**Table 1.2** Refractive index, number of atoms per unit volume (*N*/*V*), and polarizability of tellurite glasses [5]

| No | Glass | Area of glass formation metal % and halide % | Refractive index | $(N/V) \times 10^{22}$ (cm$^3$) | Polarizability ($10^{-24}$) (cm$^3$) |
|---|---|---|---|---|---|
| | $TeO_2$ (crystal) | | 2.37 | 6.78 | 7.37 |
| | $TeO_2$ (glass) | | 2.2 | 5.77 | 6.065 |
| 1 | $TeO_2$–$LiO_2$ | 12.1–34.9 | 2.108–1.93 | 6.62–7.13 | 5.78–4.78 |
| 2 | $TeO_2$–$Na_2O$ | 5.43–4.14 | 2.148–1.8 | 6.34–6.097 | 7.26–5.03 |

with $Nd^{3+}$. According to Dimitrove and Komatsu's research, former glass with a significant cationic charge has a high value for unit field strength [68].

Table 1.4 collected the obtained data of the optical basicity of neodymium NPs-doped tellurite glass series. The large value of optical basicity in the glass series revealed that the glass system is more basic. Moreover, oxide ions in the glass system act as a Lewis base which donates electrons to the surrounding cations. The high degree of donating ability of oxide ions influences strongly the acid–base characteristic. After that, Duffy proposed that the optical basicity is related to the polarization state of oxide ions in the glass system [59]. Low polarizing cations in the glass system contribute to the strong donating ability of oxide ions to the surrounding cations. Based on the previous data on oxide ion polarizability, there is good agreement with optical basicity.

The optical basicity and concentration of the glass series show a nonlinear trend. The nonlinear trend in optical basicity is caused by a structural change in the glass system. Also, it has been demonstrated that as neodymium NP concentration rises, optical basicity values do as well.

The explanation was revealed by comparing the optical basicity values of the various glass system components. Under glass former and modifier conditions, former oxides should be less basic than modifier oxides. When modifier oxide is added to the former oxide, the modification reaction transforms into an acid–base reaction where the modifier oxide ion moves toward the acidic region of the former oxide in the sequence of decreasing acidities [69].

According to the optical basicity measurements of neodymium NPs-doped tellurite glass, the addition of modifier oxide lowers the former oxide's acidity while raising the basicity of the glass system. $Nd_2O_3$, $TeO_2$, ZnO, and $B_2O_3$ each have optical basicity values of 1.014, 0.93, 0.82, and 0.425, respectively [70]. Table 1.4 lists the results for the tellurite glasses doped with neodymium NPs metallization criteria. The findings demonstrate that when concentration increases, the values of the metallization criterion drop as well. The glass samples are metalized as seen by

**Table 1.3** Electronic polarizability and oxide ion polarizability of NPs doped $\{(TeO_2)0.7(B_2O_3)0.3]0.7\ (ZnO)0.3\}1-x\ (Nd_2O_3\ NPs)\ x$, and $x$ Mol% [30]

| Electronic polarizability | (Å) Oxide ion polarizability | $\alpha o^{2-}$ (Å) |
|---|---|---|
| 0.005 | 6.091 | 2.710 |
| 0.03 | 6.272 | 2.678 |
| 0.05 | 6.655 | 2.774 |

**Table 1.4** Optical basicity and metallization criterion of $\{(TeO_2)0.7(B_2O_3)0.3]0.7\ (ZnO)0.3\}1-x\ (Nd_2O_3\ NPs)\ x$ (where $x = 0.005$, 0.03 and 0.05 Mol% [30]

| Neodymium NPs | Optical basicity, Λ | Metallization criterion |
|---|---|---|
| 0.005 | 1.191 | 0.518 |
| 0.03 | 1.175 | 0.506 |
| 0.05 | 1.170 | 0.485 |

the decline in the metallization criterion, which causes the band gap to tail off and contract [71].

The band gap of the glass system affects the metallization requirement. The valence and conduction bands have widened, according to the optical band gap's decreasing value. The glass system will become metalized as a result of this process. By contrasting their values, it is possible to determine the direct correlation between the optical band gap and the metallization criterion. The metallization requirement is lowered as the number of optical band gaps rises. The outcome complies with Dimitrov and Sakka's [72] computed metallization threshold.

## 1.4 Sellmeier Coefficients for Some Tellurite Glasses

To optimize the design characteristics of optical devices, Sellmeier coefficients are required [73]. To determine the chromatic dispersion behavior at any wavelength, notably at the optical windows of optical fiber communication systems and femtosecond technologies, these coefficients are derived for binary tellurite glasses. The measured refractive indices across the transmission range cannot be accurately represented by a single oscillator DiDomenico and Wemple's dispersion equation. To better precisely capture the data, at least a two-pole Sellmeier equation is needed. The refractive indices and their dispersion are influenced by the average electronic absorption band gap and the lattice absorption frequency, which are located in the UV and IR regions, respectively. There is a discussion of other optical characteristics. A single oscillator dispersion equation that cannot prove zero dispersion exists.

The primary component of optical fibers is silica glass ($SiO_2$), which has been widely employed to create various optical fiber amplifiers and fiber lasers with various doping. It is important to employ optical switches to process light signals without converting them to electronic form to replace the current electronic switching. These switches rely on the nonlinear optical characteristics of the materials, specifically the change in refractive index brought on by a strong optical pulse or an applied electric field. Tellurite glasses have a high figure of merit for photonic switches and a high refractive index [74]. They are effective at three of the fluoride fibers' optical windows and might even be helpful there [73]. These glasses' linear refractive index, dispersion, and chromatic dispersions are crucial properties.

Also, in reference [73], the obtained refractive indices were dispersion equation in a two-pole Sellmeier equation and Wemple's dispersion equation in a single oscillator analysis and fitting independently [2, 75, 76]. From the calculated Sellmeier coefficients, optical parameters including the Abbe number ($v_d$), average electronic absorption band gap $E_{opt}$, zero dispersion wavelength $d$, and chromatic $V$ throughout the transmission zones, 0.85, 1.33, and 1.55 m have been examined [73].

$$\text{First Sellmeier coefficients}: \; n^2 = 1.0 + B/\left(1 - C/\lambda^2\right)$$

$$\text{Second Sellmeier coefficients}: \; n^2 = A + B/\left(1 - C/\lambda^2\right) + D/\left(1 - E/\lambda^2\right)$$

where ($n$) is the refractive index, $B$ and $C$ are optical constants, and ($\lambda$) is the wavelength in μm, $C$ is the average absorption band gap, and $E \approx 7.5$ μm as in Table 1.5. Table 1.5 shows the comparison between tellurite and silicate glasses Sellmeier coefficients for tellurite and silicate glasses at 23 °C [73]. The absorption band gap for $TeO_2$ (Pure) was 4.312 eV and for $SiO_2$ was 11.8 eV. It has been found that the absorption band gap of $TeO_2$ (Pure) = 0.36 is the absorption bandgap $SiO_2$.

More research on the experimental results of specific tellurite glasses in the forms $TeO_2$–$TiO_2$–$WO_3$–$A_nO_m$, where $A_nO_m$ = $Nb_2O_5$, $Nd_2O_3$, and $Er_2O_3$, with varying mol%, has recently been reported [77]. The interpretations have been completed by obtaining the refractive index using Sellmeier coefficients and the average polarizability using the Lorentz–Lorenz equation. This means that, particles are trapped in their equilibrium positions by a linear restoring force. Consequently, a linear restoring force traps particles in their equilibrium positions. According to Goncalves et al. [78], the damping system is proportional to velocities.

A generalized Drude–Lorentz model with fractional derivative operators can be represented in the next equation:

$$\frac{d^2}{dt^2}x(t) + \gamma\frac{d^\alpha}{dt^\alpha}x(t) + \omega_0^2 x(t) = F_{\text{ext}}(t) \tag{1.11}$$

where $F_{\text{ext}} = \frac{E_0 q}{m} e^{-i\omega t}$ = the external force, $\gamma = n$—the damping constant, and $\omega_0$ is the resonance frequency. The fractional calculation allows generalizing the behavior of damped systems with the parameter of the $\alpha$ of the fractional derivative [79]. The refractive index was described as [80]:

$$n^2 = \frac{1}{2}\left[\text{Re}(\varepsilon) + \sqrt{\text{Re}(\varepsilon)^2 + \text{Im}(\varepsilon)^2}\right] \tag{1.12}$$

**Table 1.5** Sellmeier's coefficients for tellurite and silicate glasses at 23 °C [73]

| | A | B | C | D | E |
|---|---|---|---|---|---|
| $TeO_2$ | 3.548 | 0.97 | $8.266 \times 10^{-2}$ | 6.65 | 225 |
| $SiO_2$ | 1.312 | 0.79 | $1.09 \times 10^{-2}$ | 0.91 | 100 |

**Table 1.6** Sellmeier parameters for $80TeO_2$–$5TiO_2$–$(15 - x)WO_3$–xAnOm, where AnOm is $Nb_2O_5$, $Nd_2O_3$, $Er_2O_3$ and $x = 0.1;1; 3$ and 5 mol% [32]

| Sample | A | B | C | D | E |
|---|---|---|---|---|---|
| TTW-01Nb | 4.6771 | 0.0076 | 0.2612 | 76.40 | 163.84 |
| TTW-5Nb | 5.5167 | 0.0018 | 0.2390 | 366.80 | 163.84 |
| TTW-01Nd | 4.5344 | 0.0084 | 0.1235 | 148.06 | 163.84 |
| TTW-5Nd | 4.7887 | 0.0323 | 0.1200 | 212.58 | 163.84 |
| TTW-01Er | 4.70864 | 0.0019 | 0.2235 | 189.49 | 163.84 |
| TTW-5Er | 5.3342 | 0.0125 | 0.2212 | 363.12 | 163.84 |

where $Re(\varepsilon)$ is the real part, and $Im(\varepsilon)$ is the imaginary part. So, this generalized mathematical model investigates the electro-optical properties of the vitreous system [33, 36]. Data of $TeO_2$–$TiO_2$–$WO_3$–$A_nO_m$, where $A_nO_m = Nb_2O_5$, $Nd_2O_3$, and $Er_2O_3$ with different mole %, were collected in Table 1.1. So, refractive index as a function of wavelength can be obtained using the two-pole Sellmeier equation [73, 81, 82]:

$$n^2(\lambda) = A + \frac{B}{1 - \frac{C}{\lambda^2}} + \frac{D}{1 - \frac{E}{\lambda^2}} \tag{1.13}$$

The first and second terms correspond to the refractive index's contribution of greater and lower energy gaps from electronic absorption, where is the wavelength in micrometers. The final term indicates how the lattice absorption affects how much the refractive index lowers [73, 81]. The average absorption band gap ($E_{opt}$) and the $C$ coefficient are related by the relation, and $E$ can be calculated as the square of twice the IR transmission edge [81]. Table 1.6 contained the Sellmeier coefficients.

The generalized Drude–Lorentz model changes are also shown (red line). The $TeO_2$–$TiO_2$–$WO_3$–$x$Nb and $TeO_2$–$TiO_2$–$WO_3$–$x$Nd samples all exhibited the same behavior and are therefore not included in this list. All the parameters were collected in Table 1.7. An algorithm called Levenberg Marquardt was used to complete the fit. From the parameters of the constituent oxides and molar fractions, it has been anticipated that the resonance oscillator frequency ($f_o$) from the energy single oscillator may be calculated as follows [83, 84]:

$$E_0 = \sum_j x_j E_{0,j} \tag{1.14}$$

where $E_{0,j}$ = the single oscillator energy for the oxide component. While, $x_j$ = the mole fractions of each oxide in the glass. It has been found: $E_{0,TeO_2} = 6.2$ eV, $E_{0,TiO_2} = 6.2$ eV, $E_{0,WO_3} = 8.3$ eV, $E_{0,Nb_2O_5} = 6.8$ eV, $E_{0,Nd_2O_3} = 7.21$ eV, and

**Table 1.7** Adjust parameters in refractive index data obtained by Sellmeier's coefficients as a function of frequency ($f$) and the fractional derivative parameter ($\alpha$) for different damping coefficient ($\gamma$) values for $80TeO_2$–$5TiO_2$–$(15 - x)WO_3$–xAnOm, where AnOm is $Nb_2O_5$, $Nd_2O_3$, $Er_2O_3$ and $x = 0.1$; 1; 3 and 5 mol% [32]

| Sample | $f_0$ ($\times 10^{14}$ Hz) | $f_p$ ($\times 10^{14}$ Hz) | $\Gamma$ | $\alpha$ | $\varepsilon_\infty$ |
|---|---|---|---|---|---|
| TTW-01Nb | 18.21 | 1.732 ± 0.004 | 19.8 | 0.393 ± 0.006 | 4.577 ± 0.002 |
| TTW-5Nb | 16.23 | 4.344 ± 0.009 | 6 | 0.494 ± 0.01 | 5.428 ± 0.01 |
| TTW-01Nd | 18.62 | 2.461 ± 0.006 | 23.1 | 0.362 ± 0.002 | 4.455 ± 0.005 |
| TTW-5Nd | 16.78 | 3.209 ± 0.002 | 8.5 | 0.433 ± 0.005 | 4.728 ± 0.006 |
| TTW-01Er | 19.16 | 2.807 ± 0.006 | 22.3 | 0.358 ± 0.003 | 4.609 ± 0.005 |
| TTW-5Er | 16.53 | 4.292 ± 0.004 | 7.3 | 0.41 ± 0.01 | 5.240 ± 0.008 |

$E_{0,Er_2O_3} = 10.82$ eV[85–87]. Values of $E_0$ are presented in Table 1.8. It has been noticed that the $E_0$ values decrease with increasing concentration for $TeO_2$–$TiO_2$–$WO_3$–xNb and $TeO_2$–$TiO_2$–$WO_3$–xNd glasses. However, the $TeO_2$–$TiO_2$–$WO_3$–xEr glasses present the opposite behavior.

Variation of the average polarizability $X_m$ with $\alpha$, as shown by analyzing the average polarizability determined by the Lorentz–Lorenz equation [88]:

$$X_m = \frac{3}{4\pi}\frac{M}{\rho N_{AV}}\left(\frac{n^2-1}{n^2+1}\right)\left(\times 10^{24}\frac{\text{Å}^3}{\text{cm}^3}\right) \quad (1.15)$$

where: $M$ = molecular weight, $\rho$ = density, $N_{AV}$ = Avogadro's number, and $n$ = refractive index (normally taken from the sodium D line), respectively. It has been noticed that for $TeO_2$–$TiO_2$–$WO_3$–xNb set, the α values increase up to 3 mol% and remain constant at ≈ 0.49 following the $X_m$ behavior. For $TeO_2$–$TiO_2$–$WO_3$–xNd set, the α values increase up to 3 mol% and decrease to 5 mol% samples, while the $X_m$ increases up to 3 mol% and then decrease. The $TeO_2$–$TiO_2$–$WO_3$–xEr samples present the increasing behavior up to 1 mol% then a decrease for 3 mol% samples with increasing for 5 mol% samples, while the $X_m$ increases.

The degree of lattice anomaly can be correlated with the polarizability. When it comes to glasses, potential flaws brought on by their chemical makeup, such as in $TeO_2$–$TiO_2$–$WO_3$–xNb samples, might cause anomalous dispersion. When $Nb_2O_5$ is added, non-bridging oxygen (NBO) increases at the expense of bridging oxygen (BO). Since the NBO is more polarizable than the BO, an increase in mean polarizability (Xm) results in a decrease in the abnormality in the glassy system, which lowers the dispersion (lower values). Also, each dopant's unique behavior in terms of polarizability can be related to the system's mole % [28], and when the fractal cluster exhibits a high degree of optical resonance, the light-induced dipole interaction between polarizable particles can be quite potent.

**Table 1.8** $E_0$(eV) and mean polarizability $X_m$ (Å) for 80$TeO_2$–5$TiO_2$–(15 − $x$)$WO_3$–xAnOm, where AnOm is $Nb_2O_5$, $Nd_2O_3$, $Er_2O_3$ and $x$ = 0.1;1; 3 and 5 mol% [32]

| Sample | $E_0$ (eV) | $\rho$ (g/cm$^3$) | $X_m$ (Å) | $N(\times 10^{22}$ cm$^{-3})$ | $\phi(\times 10^{-29})$ | $N\phi$ ($\times 10^{-7}$ cm$^{-3}$) |
|---|---|---|---|---|---|---|
| TTW-01Nb | 6.51 | 5.455 | 6.452 | 1.973 | 4.287 | 8.461 |
| TTW-5Nb | 6.44 | 5.673 | 6.651 | 2.032 | 0.662 | 1.345 |
| TTW-01Nd | 6,51 | 5.214 | 6.698 | 1.885 | 2.223 | 4.191 |
| TTW-5Nd | 6.46 | 5.680 | 6.795 | 1.992 | 1.237 | 2.465 |
| TTW-01Er | 6.52 | 5.778 | 5.669 | 2.089 | 1.542 | 3.221 |
| TTW-5Er | 6.64 | 5.972 | 6.358 | 2.067 | 0.667 | 1.378 |

## 1.5 Nano-Rare-Earth Ions in Tellurite Glasses

Additionally, optical characteristics between two glass series (*a*—ordinary glass and *b*—glass with nanoparticles) have been researched in 2017 [28]. The two glass series (*a*—ordinary glass and *b*—glass with nanoparticles) were effectively made with compositions of $[(TeO_2)0.70\ (B_2O_3)0.30]0.70\ (ZnO)0.301 - y\ (Er_2O_3/Er_2O_3$ nanoparticles)$y$ and ($y$) varied from 0.005 to 0.05 mol%.

Measured refractive indices for the two glass series (*a*—regular glass and *b*—glass containing nanoparticles) are listed in Table 1.1. Nanoparticle-infused glass has a higher density than regular glass, according to research. The concentration of erbium and erbium nanoparticles oxide, as well as the refractive index of the glass series, are growing linearly. Glasses containing nanoparticles have a higher refractive index than regular glass (*a*).

It is possible to explain the change in refractive index.

(a) density,
(b) non-bridging oxygen,
(c) polarizability, and
(d) coordinate number.

The structure of ZBTSm-NPs glasses is more compact than ZBTSm-MPs glasses, which accounts for its high refractive index value. Therefore, tiny particles condense the gaps between atoms and restrict the spread of photon energy in the glass network [25]. Moreover, the refractive index trend and dopant concentrations are similar in both the ZBTSm-MPs and ZBTSm-NPs glass series. The change in refractive index is impacted by the rare-earth ions because they increase the amount of non-bridging oxygen in the glass structure [25].

According to Greenwood's [89] theory, the uniform distribution of the smaller particles throughout the bulk materials is what causes their high solubility. Additionally, Toy et al. [90] investigated how particle size affected the margination of nanoparticles and discovered that as particle size declines, density rises. The cohesion of atoms and the volume of unit cells are the two parameters that control the size-dependent density of nanoparticles.

A previous study by Nanda et al. [91] stated that the cohesive energy decreases with the size of nanoparticles. While the lattice constant of the nanoparticle may increase or decrease. The glass with nanoparticles will offer a potential material for nano-photonic devices.

## 1.6 Refractive Index in Terahertz (THz)

Values of both refractive index and absorption coefficients $n(\nu)$ and $\alpha(\nu)$ in the terahertz (THz) bandwidth have been reported [92]. Storm et al. [93], Storm and Taylor [94] and Schlomann [93] investigated refractive index and absorption coefficients using the next relation:

$$n(\nu)\,\alpha(\nu) = K \times f^{\beta}, \tag{1.16}$$

Both of the glass constant, $K$ determined and the power $\beta$, depend on the glass composition as in the next relation;

$$K = e^{2*}Nk^2/h^2\rho c V_D^3 \tag{1.17}$$

where $N$, $(e*)$, $k$, $\rho$, $c$, and $V_D$ are density of charge fluctuations of amplitude, the local field correction factor of amplitude $(e^*)$, $h$ reduced Planck constant, density, speed of light in a vacuum, and the Debye velocity of sound, respectively.

Tellurite glasses $53TeO_2$–$25WO_3$–$22Na_2O$ had the highest refractive index (4.76) and absorption coefficients (44.06 $cm^{-1}$) at terahertz. Moreover, tellurite glasses had the heights $K$ (715 $cm^{-1}$) and $\beta$ (1.74 $cm^{-1}$), respectively.

It can be concluded that refractive indices of tellurite in both visible and terahertz (THz) frequencies are higher than those of other glasses. This investigation consists of studying glasses, viz.:

- Experimental measurement of the linear refractive index showed that tellurite glasses with higher density produced a higher refractive index. Moreover, tellurite glasses doped with nanoparticles produced tellurite glasses with a refractive index higher than the ordinary glass.
- The dependence of the refractive index of glasses on temperature has been discussed based on the thermo-optic coefficient. Also, $TeO_2$-based glasses got high potential.
- Theoretical interpretation for linear refractive index had been achieved by using the definition of electronic polarizability, molar refraction, number of polarizable atoms/unit volume, optical basicity, and metallization criterion.
- Sellmeier coefficients evaluated Abbe number, average electronic absorption band gap, zero dispersion wavelength, and chromatic $V$ through both first and second Sellmeier coefficients.
- The linear refractive index of tellurite glasses contained increased due to the modification with erbium nanoparticles oxide.
- Tellurite glasses had the heights absorption coefficient and fitting parameters 44.06 ($cm^{-1}$) and 1.74 ($cm^{-1}$), respectively. Besides the above, it is important to mention the research work on thermal, elastic & mechanical and electrical properties of tellurite glasses and others has been achieved [5, 66].

## References


1. Stanworth, J. 1952. Tellurite glasses. *Nature* 169: 581–582.
2. Ivanova, Y. 1982. Phase equilibria in the MnO–$TeO_2$ system. *Materials Chemistry* 7 (4): 449–453.
3. Flynn, B., A. Owen, and J. Robertson. 1977. Electronic conduction in vanadium tellurite glasses. In *Proceedings 7th international conference on amorphous liquid and semiconductors*.

4. Lambson, E., et al. 1984. The elastic behavior of $TeO_2$ glass under uniaxial and hydrostatic pressure. *Journal of Non-Crystalline Solids* 69 (1): 117–133.
5. El-Mallawany, R.A. 2011. *Tellurite glasses handbook: Physical properties and data*, 2nd ed., 532. New York: CRC-Press.
6. El-Mallawany, R. 2017. Introduction to tellurite glasses. In *Technological advances in tellurite glasses: Properties, processing, and applications*, 1–13.
7. El-Mallawany, R. 2018. *Editor Tellurite glass smart materials: Applications in optics and beyond*, 1–297. New York: Springer International Publishing. https://doi.org/10.1007/978-3-319-76568-6
8. El-Mallawany, R. 2018. Some physical properties of tellurite glasses. In: *Tellurite glass smart materials: Applications in optics and beyond*, 1–16
9. Jha, A., et al. 2012. Rare-earth ion doped $TeO_2$ and $GeO_2$ glasses as laser materials. *Progress in Materials Science* 57 (8): 1426–1491.
10. El-Mallawany, R. 2020. Tellurite glasses, properties, structure and applications. *Preface Journal of Non-Crystalline Solids Special Issue* 528: 119784.
11. Hossain, M.K., et al. 2021. A review on optical applications, prospects, and challenges of rare-earth oxides. *ACS Applied Electronic Materials* 3 (9): 3715–3746.
12. Patra, P., and K. Annapurna. 2022. Transparent tellurite glass-ceramics for photonics applications: A comprehensive review on crystalline phases and crystallization mechanisms. *Progress in Materials Science* 125: 100890.
13. Raghvender, R., et al. 2022. A Buckingham interatomic potential for thallium oxide ($Tl_2O$): Application to the case of thallium tellurite glasses. *Computational Materials Science* 201: 110891.
14. Ersundu, M.Ç., et al. 2019. Physical, mechanical and gamma-ray shielding properties of highly transparent $ZnO–MoO_3–TeO_2$ glasses. *Journal of Non-Crystalline Solids* 524: 119648.
15. Mallur, S.B., and P.K. Babu. 2022. Optical properties of Praseodymium ($Pr^{3+}$) doped bismuth boro-tellurite glasses containing CdSe nanoparticles. *Materials Research Bulletin* 147: 111651.
16. He, J., et al. 2022. Novel 3.1 μm and enhanced 2.7 μm emissions in Er3+ doped fluorotellurite glasses ceramic. *Journal of Alloys and Compounds* 895: 162606.
17. Machado, T.M., et al. 2022. Copper nanoparticles—containing tellurite glasses: An efficient SERS substrate. *Materials Chemistry and Physics* 278: 125597.
18. Aouaini, F., et al. 2022. Visible to infrared down conversion of $Er^{3+}$ doped tellurite glass for luminescent solar converters. *Journal of Alloys and Compounds* 894: 162506.
19. Belançon, M.P., et al. 2022. Float, borosilicate and tellurites as cover glasses in Si photovoltaics: Optical properties and performances under sunlight. *Journal of Physics and Chemistry of Solids* 161: 110396.
20. Marek, Ł, and M. Sobczyk. 2022. Optical properties, concentration and thermal quenching of luminescence of $Dy^{3+}$-doped $La_2O_3–Na_2O–ZnO–TeO_2$ glasses. *Journal of Non-Crystalline Solids* 576: 121238.
21. Ferodolin, I., et al. 2022. Plasmonic effect of bimetallic $TiO_2/Al_2O_3$ nanoparticles in tellurite glass for surface-enhanced Raman scattering applications. *Journal of Luminescence* 241: 118488.
22. Marzuki, A., F.D. Ega, and A. Saraswati. 2022. Effect of $B_2O_3$ addition on thermal and optical properties of $TeO_2–ZnO–Bi_2O_3–TiO_2$ glasses. *Materials Research Express* 9 (2): 025203.
23. Strutynski, C., et al. 2022. Physicochemical properties and fiber-drawing ability of tellurite glasses in the $TeO_2–ZnO–Y_2O_3$ ternary system. *Materials* 15 (3): 1177.
24. Alekseev, R., et al. 2022. Local atomic structure of the high refractive index $La_2O_3–Nb_2O_5–B_2O_3$ glasses. *Journal of Alloys and Compounds* 917: 165357.
25. Azlan, M., et al. 2021. Comprehensive comparison on optical properties of samarium oxide (micro/nano) particles doped tellurite glass for optoelectronics applications. *Journal of Materials Science: Materials in Electronics* 32 (11): 14174–14185.
26. Al-Qahtani, S.D., et al. 2022. Production of smart nanocomposite for glass coating toward photochromic and long-persistent photoluminescent smart windows. *Ceramics International* 48 (1): 903–912.

27. Chandrappa, V., et al. 2022. Spectral studies of $Dy^{3+}$: Zincphosphate glasses for white light source emission applications: A comparative study. *Journal of Non-Crystalline Solids* 583: 121466.
28. Azlan, M., et al. 2017. Optical properties of zinc borotellurite glass system doped with erbium and erbium nanoparticles for photonic applications. *Journal of Materials Science: Materials in Electronics* 28: 4318–4327.
29. Azlan, M., et al. 2018. Linear and nonlinear optical efficiency of novel neodymium nanoparticles doped tellurite glass for advanced laser glass. *Education of JSMT* 5 (2): 47–66.
30. Azlan, M., et al. 2019. Electronic polarizability and third-order nonlinearity of $Nd^{3+}$ doped borotellurite glass for potential optical fiber. *Materials Chemistry and Physics* 236: 121812.
31. Azlan, M., et al. 2019. Upconversion properties of erbium nanoparticles doped tellurite glasses for high efficient laser glass. *Optics Communications* 448: 82–88.
32. Azlina, Y., et al. 2020. Optical performance of neodymium nanoparticles doped tellurite glasses. *Physica B: Condensed Matter* 577: 411784.
33. Jan, N.M., et al. 2019. Thermal, structural and magnetic properties of $TeO_2$–MgO–$Na_2O$–$Nd_2O_3$ glass system with NiO nanoparticles. *Journal of Non-Crystalline Solids* 522: 119566.
34. Sales, T.O., C. Jacinto, W.F. Silva, R. Antunes, D.T. Dias, A. Gonçalves, R. El-Mallawany, N.G.C. Astrath, and A. Novatski. 2022. White light source and optical thermometry based on zinc-tellurite glass tri-doped with $Tm^{3+}$/$Er^{3+}$/$Sm^{3+}$. *Journal Alloys and Compounds* 899: 163305.
35. Song, C., et al. 2022. Structure, and luminescence properties in $Tm^{3+}$/$Yb^{3+}$ co-doped bismuth-tellurite glass for 2 μm fiber lasers. *Journal of Non-Crystalline Solids* 593: 121784.
36. Desirena, H., et al. 2022. $Eu^{3+}$ heavily doped tellurite glass ceramic as an efficient red phosphor for white LED. *Journal of Luminescence* 250: 119080.
37. Alkallas, F.H., et al. 2022. Enhancement of spectroscopic properties of Er3+ doped TeO2–Nb2O5–ZnO glassy systems for optical amplifiers at 1.53 μm. *Optik* 265: 169507.
38. Jiang, X., et al. 2022. Spectroscopic properties and Stark splitting of $Nd^{3+}$ ions in different host glasses to obtain pure blue laser. *Journal of Luminescence* 249: 119005.
39. Biswas, J., et al. 2022. Optical and luminescence properties of $Sm_2O_3$ doped SrO–PbO–ZnO–$P_2O_5$–$TeO_2$ glasses for visible laser applications. *Solid State Sciences* 129: 106910.
40. Luo, J., et al. 2022. Effect of fluorine substitution on the structure and spectral property of fluorotellurite glass for upconversion luminescence thermometry. *Journal of Luminescence* 247: 118906.
41. Feng, S., et al. 2022. Intense 2.71-μm fluorescence emission in low hydroxyl heavily $Er^{3+}$-doped fluorotellurite glass for mid-infrared fiber laser. *Journal of Non-Crystalline Solids* 586: 121569.
42. Awshah, A.A.A., et al. 2022. Polarizability and optical properties of $TeO_2$–ZnO glass system doped with $Nd_2O_3$. *Journal of Materials Science: Materials in Electronics* 33 (17): 13493–13505.
43. Teresa, P.E., et al. 2022. Study on the luminescence behavior of $Dy^{3+}$ ions activated, modifier dependent alkali boro-tellurite glasses for white LED application. *Optik* 259: 169024.
44. Yuan, J., et al. 2022. Strong and broadband 2.7 μm fluorescence of $Yb^{3+}$/$Er^{3+}$ co-doped ZnO modified tungsten tellurite glasses for broadband mid-infrared optical amplifiers and tunable fiber lasers. *Journal of Luminescence* 246: 118840.
45. Andrianov, A.V., and E.A. Anashkina. 2022. Thermo-optical control of L-band lasing in Er-doped tellurite glass microsphere with blue laser diode. *Optics Letters* 47 (9): 2182–2185.
46. Zhou, D., et al. 2022. Spectroscopic properties of $Er^{3+}$-doped lithium-modified bismuth tellurite glasses for broadband near-infrared emission. *Journal of Non-Crystalline Solids* 583: 121468.
47. Wang, J., et al. 2022. Thulium-doped fluorotellurite glass fibers for broadband S-band amplifiers. *Optics Letters* 47 (8): 1964–1967.
48. Niu, L., et al. 2022. Optimized tellurite glasses containing $CsPbBr_3$-quantum dots for white-light emitting diodes. *Journal of Non-Crystalline Solids* 581: 121429.
49. Ding, J., et al. 2022. $Pr^{3+}$/$Tm^{3+}$/$Er^{3+}$ tri-doped tellurite glass with ultra-broadband luminescence in the optical communication band. *Ceramics International* 48 (6): 8779–8782.

50. Amin, R., et al. 2022. Tellurite glass based optical fiber for the investigation of supercontinuum generation and nonlinear properties. *Physica Scripta* 97 (3): 030007.
51. Burtan-Gwizdala, B., et al. 2022. Strong emission at 1000 nm from $Pr^{3+}/Yb^{3+}$-codoped multicomponent tellurite glass. *Pure and Applied Chemistry* 94 (2): 147–156.
52. Lesniak, M., et al. 2022. Tellurite-germanate ZnTe: Glass-ceramics doped with $EuF_3$ for optical fiber application. In: *Fiber lasers and glass photonics: Materials through applications III*. SPIE.
53. Jayasimhadri, M., and D. Haranath. 2022. Spectroscopic investigations of $Dy^{3+}$-doped tungstate–tellurite glasses for solid-state lighting applications. *International Journal of Applied Glass Science* 4: 645–654.
54. Boonin, K., et al. 2022. Fabrication and characterization of $Eu^{3+}$ doped tellurite glasses for laser materials. *Integrated Ferroelectrics* 225 (1): 199–211.
55. Barczak, K., et al. 2022. Measurements of Verdet constant in heavy metal oxide glasses for magneto-optic fiber current sensors. *Optical Materials* 123: 111942.
56. Wongwan, W., et al. 2022. Structural and luminescence characterizations of $Tb^{3+}$ ion doped boro-tellurite glasses for LED applications. *Integrated Ferroelectrics* 224 (1): 62–70.
57. Awshah, A., et al. 2023. Novel of neodymium nanoparticles zinc tellurite glasses in experimental and theoretical elastic properties using artificial intelligence approach. *Chinese Journal of Physics* 81: 332–353.
58. D'Souza, A.N., et al. 2023. $TeO_2$ for enhancing structural, mechanical, optical, gamma and neutron radiation shielding performance of bismuth borosilicate glasses. *Materials Chemistry and Physics* 293: 126657.
59. Duffy, J.A. 1989. Optical basicity of titanium(IV) oxide and zirconium(IV) oxide. *Journal of the American Ceramic Society* 72 (10): 2012–2013.
60. El-Mallawany, R., M.D. Abdalla, and I.A. Ahmed. 2008. New tellurite glass: Optical properties. *Materials Chemistry and Physics* 109 (2–3): 291–296.
61. Honma, T., et al. 2010. Thermo-optic properties and electronic polarizability in alkali tellurite glasses. *Journal of the American Ceramic Society* 93 (10): 3223–3229.
62. Azlan, M., et al. 2020. Effect of $Gd^{3+}$ on optical and thermal properties of tellurite glass. *Journal of Theoretical and Applied Physics* 14: 137–147.
63. Krishna, V.M., et al. 2022. Near-IR luminescence in $Nd^{3+}$ ions doped $Na_2O$–$BaF_2$–$CaF_2$–$B_2O_3$–$TeO_2$ glasses for 1064 nm laser and fiber amplifier applications. *Journal of Non-Crystalline Solids* 590: 121671.
64. Ding, J., et al. 2022. A new $Nd^{3+}/Tm^{3+}/Ho^{3+}$ tri-doped tellurite glass for multifunctional applications. *Materials Research Bulletin* 150: 111777.
65. El-Mallawany, R. 1992. The optical properties of tellurite glasses. *Journal of applied physics* 72 (5): 1774–1777.
66. El-Mallawany, R.A. 2002. *Tellurite glasses handbook: Physical properties and data*. New York: CRC Press.
67. Azlan, M., et al. 2018. Linear and nonlinear optical efficiency of novel neodymium nanoparticles doped tellurite glass for advanced laser glass. *Educational and JSMT* 5 (2): 47–66.
68. Dimitrov, V., and T. Komatsu. 2010. An interpretation of optical properties of oxides and oxide glasses in terms of the electronic ion polarizability and average single bond strength. *Journal of University in Chemical Technology of Metallaurgy* 5 (3): 219–250.
69. Sidek, H., et al. 2004. Formation and elastic behavior of lead-magnesium chlorophosphate glasses. *Turkish Journal of Physics* 28 (1): 65–72.
70. Dimitrov, V., and T. Komatsu. 2005. Classification of oxide glasses: A polarizability approach. *Journal of Solid State Chemistry* 178 (3): 831–846.
71. Sun, J., et al. 2011. Glass formation and properties of Ge–Te–$BiI_3$ far infrared transmitting chalcohalide glasses. *Spectrochimica Acta Part A: Molecular and Biomolecular Spectroscopy* 79 (5): 904–908.
72. Dimitrov, V., and S. Sakka. 1996. Linear and nonlinear optical properties of simple oxides. *II. Journal of Applied Physics* 79 (3): 1741–1745.

73. Ghosh, G. 1995. Sellmeier coefficients and chromatic dispersions for some tellurite glasses. *Journal of the American Ceramic Society* 78 (10): 2828–2830.
74. Lines, M. 1991. Oxide glasses for fast photonic switching: A comparative study. *Journal of Applied Physics* 69 (10): 6876–6884.
75. DiDomenico, M. 1972. Material dispersion in optical fiber waveguides. *Applied Optics* 11 (3): 652–654.
76. Wemple, S. 1977. Optical oscillator strengths and excitation energies in solids, liquids, and molecules. *The Journal of Chemical Physics* 67 (5): 2151–2168.
77. El-Mallawany, R., et al. 2021. Refractive index behavior of tellurite glasses. *Optical Materials* 112: 110810.
78. Gonçalves, A., et al. 2019. A generalized Drude-Lorentz model for refractive index behavior of tellurite glasses. *Journal of Materials Science: Materials in Electronics* 30: 16949–16955.
79. Evangelista, L.R., and E.K. Lenzi. 2018. *Fractional diffusion equations and anomalous diffusion*. Cambridge: Cambridge University Press.
80. Reitz, J.R. 2009. *Foundations of electromagnetic theory*. Noida: Pearson Education India.
81. Ghosh, G. 1997. Sellmeier coefficients and dispersion of thermo-optic coefficients for some optical glasses. *Applied Optics* 36 (7): 1540–1546.
82. Wang, T.-J., et al. 2007. Sellmeier equations for green, yellow, and orange colored Hg Ga 2 S 4 crystals. *Applied Physics Letters* 90 (18): 181913.
83. Brown, R.N. 1987. Material dispersion in heavy metal oxide glasses containing $Bi_2O_3$. *Journal of Non-Crystalline Solids* 92 (1): 89–94.
84. Shannon, R., et al. 2002. Refractive index and dispersion of fluorides and oxides. *Journal of Physical and Chemical Reference Data* 31 (4): 931–970.
85. Takebe, H., S. Pujino, and K. Morinaga. 1994. Refractive-index dispersion of tellurite glasses in the region from 0.40 to 1.71 μm. *Journal of the American Ceramic Society* 77 (9): 2455–2457.
86. El-Mallawany, R. 2014. Evaluation of optical parameters of some tellurite glasses. *Optik* 125 (20): 6344–6346.
87. Butera, S., and M. Di Paola. 2014. A physically based connection between fractional calculus and fractal geometry. *Annals of Physics* 350: 146–158.
88. Shalaev, V.M., R. Botet, and R. Jullien. 1991. Resonant light scattering by fractal clusters. *Physical Review B* 44 (22): 12216.
89. Greenwood, G. 1956. The growth of dispersed precipitates in solutions. *Acta Metallurgica* 4 (3): 243–248.
90. Toy, R., et al. 2011. The effects of particle size, density and shape on margination of nanoparticles in microcirculation. *Nanotechnology* 22 (11): 115101.
91. Nanda, K., et al. 2015. Study of vibrational spectroscopy, linear and non-linear optical properties of $Sm^{3+}$ ions doped $BaO–ZnO–B_2O_3$ glasses. *Solid State Sciences* 45: 15–22.
92. Tostanoski, N.J., and S.K. Sundaram. 2023. Universal power-law of terahertz optical properties of borosilicate, tellurite, and chalcogenide glass families. *Scientific Reports* 13 (1): 2260.
93. Schlömann, E. 1964. Dielectric losses in ionic crystals with disordered charge distributions. *Physical Review* 135: A413.
94. Strom, U., J. Hendrickson, R. Wagner, and P. Taylor. 1974. Disorder-induced far infrared absorption in amorphous materials. *Solid State Communication* 15: 1871–1875.
95. Strom, U., and P. Taylor. 1977. Temperature and frequency dependences of the far-infrared and microwave optical absorption in amorphous materials. *Physical Review B Condensed in Matter* 16: 5512.

# Chapter 2
# Nonlinear Refractive Index of Some Tellurite Glasses

The current chapter, which is broken up into five subsections, discusses the nonlinear refractive index of tellurite glasses:

1. measurement of nonlinear refractive index,
2. relationship between nonlinear and linear refractive indices,
3. models of nonlinear refractive index,
4. nonlinear optical absorption,
5. some data of nonlinear refractive index and effect of NPs.

In nonlinear optics, it is essential to know:

What does "nonlinear optics" mean? The subject of nonlinear optics focuses on how powerful light affects material properties.
The optical response of a material usually scales linearly with the amplitude of the electric field. However, the material properties can alter more quickly at high powers.
What characteristics are nonlinear optical? The area of optics known as nonlinear optics (NLO) studies how light behaves in nonlinear media. The polarization density $P$ in this medium reacts nonlinearly to the light's electric field ($E$).

## 2.1 Measurement of Nonlinear Refractive Index

The nonlinear index $n_2$ in nonlinear optics is measured using the $z$-scan method, as shown in Fig. 2.1. Since nonlinear absorption may have an impact on how the nonlinear index is measured. More than that, the closed technique and open approach are frequently combined to rectify the calculated result. The closed-aperture version of the $z$-scan setup is used to measure the actual portion of the nonlinear refractive index.

 
R. El-Mallawany, *The Physics of Advanced Optical Materials: Tellurite Glasses*,
https://doi.org/10.1007/978-3-031-45245-1_2

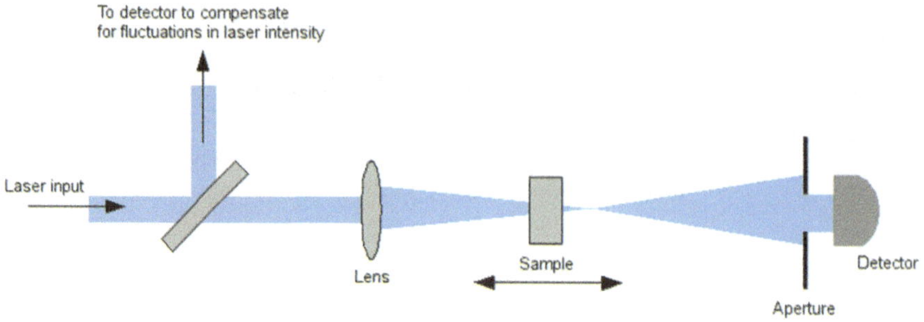


**Fig. 2.1** Schematic of *Z*-scan setup

Due to the nonlinear material's behavior in this form, which resembles a weak *z*-dependent lens, the far-field aperture allows for the identification of the slight beam distortions in the original beam. The imaginary part of the nonlinear refractive index or the nonlinear absorption coefficient is measured using the *z*-scan setup in its open aperture form. Additionally, the full signal is measured by the detector during open aperture measurements rather than the far-field aperture. Small beam distortions become unimportant when measuring the full signal, and the *z*-dependent signal change is solely caused by nonlinear absorption. When the laser intensity at a certain place in the medium does not alone control the laser's nonlinear response; it also depends on the laser intensity in the vicinity. A nonlocal nonlinear optical response will be used to describe it.

In general, a number of mechanisms, some of which may be nonlocal, may contribute to the nonlinearity. For instance, when the nonlinear medium is dispersed inside a dielectric solution, the optical field action causes nonlocal spatial reorientation of the dipoles (permanent or induced molecular dipoles), which alters the electric field that the nonlinear medium experiences.

## 2.2 Relationship Between Linear and Nonlinear Refractive Indices

In dielectric materials, the bound particles are held together by elastic connections created by the collisions of charged particles. Positive charges (ions) are attracted and slightly repelled by negative charges (electrons) when an electric field is applied across the substance. As a result, a group of induced dipole moments are created, leading to polarization.

The electric field can take many different forms, such as a voltage across the material, light passing through it, or a combination of the two. The electric and magnetic fields that make up a light wave fluctuate sinusoidally at "optical" frequencies (= $10^{13}$–$10^{17}$ Hz). Charged particles move oscillatory and form oscillating dipoles when a light beam passes through a dielectric medium. Since the impact of the electric field is so much greater than that of the magnetic field, the magnetic field can be ignored.

Also, due to their smaller mass than ions, electrons' mobility is important at high optical frequencies (the ultraviolet and visible portions of the spectrum). The wave's path through the medium is altered by the electric dipoles, which oscillate at the same frequency as the incident electric field. The electric displacement is:

$$D = \varepsilon_o E + P \tag{2.1}$$

where ($\varepsilon_o$) = the free space permittivity and the polarization, ($P$) can be expressed in terms of the electric field ($E$), and the susceptibility ($\chi$)

$$P = \varepsilon_o\{\chi^{(1)}E + \chi^{(2)}E + \chi^{(3)}E + \cdots\} \tag{2.2}$$

where $\chi^{(1)}$ = the linear electric susceptibility, $\chi^{(2)}$ = the quadratic nonlinear susceptibility, and $\chi^{(3)}$ = the cubic nonlinear susceptibility and the field-dependent refractive index can be written as:

$$n^2 = 1 + \chi^{(1)} + \chi^{(2)} + \cdots \tag{2.3}$$

The induced dipole moment will oscillate as a result of the electromagnetic wave's oscillation, which will generate light with a different frequency than the entering light wave. It should be emphasized that the applied electric field, which is the total field employed in the equation above, might include both electric and optical contributions.

Nonlinear optical processes in glasses and glass-based composites were described in Simmons and El-Bayoumim [1]. The applications of susceptibility $\chi^{(1)}$ by refraction in optical fibers, lenses, and physical optics. While the application of second susceptibility $\chi^{(2)}$ by frequency doubling is in the harmonic generation.

The refractive index of a material subject to an intense electrical field may be described by

$$n = n_o + \Delta n \tag{2.4}$$

where $n_o$ is the linear refractive index, and $\Delta n$ is the intensity-dependent index change. For a light beam of electric field $E$, the index change is given by:

$$\Delta n = n_2 E^2$$

where $n_2$ is the nonlinear refractive index. While the nuclear contribution is contained in the imaginary part of $\chi^{(3)}$ and is of a resonant type, the electronic contribution of $n_2$ is included in the real part of $\chi^{(3)}$ and is of a no-resonant type. The optical frequencies are typically out of proportion to a material's vibrational frequencies. Re $\chi^{(3)} > >$ Im $\chi^{(3)}$ applies here. Next, the nonlinear refractive index for linearly polarized monochromatic light in an isotropic medium or a cubic crystal can be stated by the following equation,

$$n_2 = (12\pi/n_o)\mathrm{Re} \cdot \chi^{(3)} \tag{2.5}$$

Simple inorganic oxides are frequently the first ingredients in the production of optical materials that are crystalline and amorphous. It is therefore of great importance to understand how their linear and nonlinear optical characteristics interact.

## 2.3 Models of Nonlinear Refractive Index

Two models have been put out to describe the characteristics of the nonlinear refractive index and the third-order nonlinear optical susceptibility. In the year 1978, Boling et al. [2] deduced the formula for calculating $n_2$ based on refractive index $n_o$ and Abbe number $\nu_d$. A suitable relationship between the linear refractive index $n_0$, the energy gap $E_{opt}$, and the nonlinear refractive index $n_2$ has been looked for based on experimental and theoretical data reported in the literature for numerous simple oxides.

The two characteristics under examination, linear refractive index and energy gap, $E_{opt}$ of the solids, are crucial for characterizing the nonlinear refractive index, according to analysis of the various models discussed above.

According to Duffy [3], the energy gap of oxides and its molar refraction exhibits a strong correlation:

$$E_{opt} = 20(1 - R_m/V_m)^2 \tag{2.6}$$

Using the Lorentz–Lorentz equation,

$$1 - \{1 - R_m/V_m\} = \left(E_g/20\right)^{0.5} \tag{2.7}$$

where $n_o$ = the linear refractive index. This equation shows the possibility of calculating the metallization criterion $(1 - R_m/V_m)$ of an oxide based on its linear refractive index or energy gap, respectively.

It is known that as the linear refractive index rises and the energy gap of the oxides decreases, the nonlinear refractive index rises as well. This is connected to the oxides' rising metallicity. According to Dimitrove and Sakka [4], metallization criteria for oxides with high nonlinear refractive indices range from 0.3 to 0.4.

## 2.4 Nonlinear Optical Absorption

The nonlinear optical absorption has been measured for $(TeO_2)$–$(B_2O_3)$–(ZnO) glasses doped with $(Nd_2O_3)$ [5]. To calculate the nonlinear optical absorption coefficient, the following equation was used to fit the curves:

$$T(z) = \frac{\ln[1 + q_0(z)]}{q_0(z)} \tag{2.8}$$

It has been reported that the valence electrons in the non-bridging oxygen are less stable and weakly bound to the host material [6].

The nonlinear absorption coefficient was calculated to be $7.49 \times 10^{-3}$. The third-order susceptibility of $Nd^{3+}$ is calculated by applying ($\eta_2$) and ($\beta$) values as shown in the following equation;

$$|\chi^{(3)}| = \left[\left(\mathrm{Re}\chi^{(3)}\right)^2 + \left(\mathrm{Im}\chi^{(3)}\right)^2\right]^{1/2} \tag{2.9}$$

where $\mathrm{Re}(\chi^3)$ and $\mathrm{Im}(\chi^3)$ are the real and imaginary parts of the third-order susceptibility. It was discovered that the third-order susceptibility of borotellurite glasses doped with $Nd^{3+}$ varied nonlinearly. The third-order nonlinearity had a substantial correlation with the glass network's structural features.

## 2.5 Data of Nonlinear Refractive Index of Some Tellurite Glasses

The glass $[(TeO_2)0.70\ (B_2O_3)0.30]0.7\ (ZnO)0.31 - y\ (Nd_2O_3)y$, where $y = 0.005$, 0.01, 0.02, 0.03, 0.04, and 0.05 mol%, have had their nonlinear refractive index [5]. The CuO-doped $TeO_2$–$B_2O_3$ glass system was explored by Umar et al. [7] for its optical basicity, polarizability, and spectroscopic investigations. The linear and nonlinear optical efficiency of innovative neodymium nanoparticles doped tellurite glass for advanced laser glass was also determined by Azlan et al. [8]. Neodymium nanoparticles (NPs)-doped tellurite glass with a chemical composition of neodymium nanoparticles $[(TeO_2)0.70\ (B_2O_3)0.30]0.7\ (ZnO)0.31 - y\ (Nd_2O_3)y$, where $y =$ 0.005, 0.01, 0.02, 0.03, 0.04, and 0.05 mol%, as in Table 2.1 exhibit an optical basicity trend along with concentration.

In the current work, neodymium NPs were added to tellurite glass, and the nonlinear refraction coefficient, 0 ($cm^2/W$), was found. It is clear that the Z-scan curves display a peak-to-valley ratio that is consistent with nonlinear refraction that is positive and self-focusing behavior. Along with the dopant concentration, Table 2.1 also included the nonlinear refractive index coefficient for neodymium NPs in nonlinear variation. The influence of dopant concentration may be the cause

**Table 2.1** Nonlinear optical properties of some tellurite glasses

| Glass composition | Reference | $n_2$ ($\times 10^{-12}$) | No | $X$ ($\times 10^{-6}$) | $E_{opt}$ (eV) | Λ | $M$ |
|---|---|---|---|---|---|---|---|
| {[($TeO_2$)0.70 ($B_2O_3$)0.30]0.7 (ZnO)0.3}1 − $y$ ($Nd_2O_3$)$y$, $Y$ = mol% | [5] | | | | | | |
| 0.005 | | 2.35 | 1.76 | 5.06 | 3.331 | 1.401 | 0.589 |
| 0.01 | | 1.29 | 1.8 | 10.2 | 3.323 | 1.400 | 0.574 |
| 0.02 | | 2.20 | 1.81 | 4.99 | 3.320 | 1.399 | 0.568 |
| 0.03 | | 1.00 | 1.83 | 6.71 | 3.311 | 1.397 | 0.560 |
| 0.04 | | 4.15 | 1.84 | 27.3 | 3.310 | 1.396 | 0.555 |
| 0.05 | | 0.633 | 1.86 | 18.0 | 3.307 | 1.173 | 0.549 |
| (1x)[($B_2O_3$)($TeO_2$)0.7]–$x$(CuO) | [7] | | | | | | |
| $X = 0$ | | | 2.318 | | | | |
| $X = 0.005$ | | | 2.321 | | | | |
| $X = 0.01$ | | | 2.345 | | | | |
| $X = 0.15$ | | | 2.357 | | | | |
| $X = 0.02$ | | | 2.378 | | | | |
| ($TeO_2$)0.7($B_2O_3$)0.3]0.7 (ZnO)0.3}1 − $x$ ($Nd_2O_3$ NPs), $x$ = mol % $Nd_2O_3$ NPs (with a size of particles ~ 15 nm) | [8] | | | | | | |
| 0.005 | | 0.734 | | 14.6 | 3.209 | | 0.518 |
| 0.01 | | 3.14 | | 10.0 | 3.201 | | 0.517 |
| 0.02 | | 1.63 | | 8.56 | 3.198 | | 0.514 |
| 0.03 | | 1.78 | | 8.53 | 3.193 | | 0.506 |
| 0.04 | | 2.98 | | 33.8 | 3.179 | | 0.501 |
| 0.05 | | 1.91 | | 34.3 | 3.178 | | 0.485 |
| ($TeO_2$)0.7($B_2O_3$)0.3]0.7 (ZnO) 0.3}1 − $x$ ($Nd_2O_3$ NPs)$x$, $x$ = mol%, NPs ($Nd_2O_3$, 20–30 nm) | [19] | | | | | | |
| 0.005 | | 0.734 | | 14.6 | 3.209 | | |
| 0.01 | | 3.14 | | 10.0 | 3.201 | | |
| 0.02 | | 1.63 | | 8.56 | 3.198 | | |
| 0.03 | | 1.78 | | 8.53 | 3.193 | | |
| 0.04 | | 2.98 | | 33.8 | 3.179 | | |
| 0.05 | | 1.91 | | 34.3 | 3.178 | | |
| Silica | [20] | | 1.45 | $2.0 \times 10^{-22}$ | | | |
| Chalcogenides | [20] | | 2.35 | 1251 | | | |

of this difference. Neodymium NPs' nonlinear refractive index is also quite valuable. The linear refractive index and nonlinear refractive index have a relationship. A glass substance with a high-refractive index also has a high nonlinear refractive index, according to Chen et al. [9].

Neodymium NPs have a nonlinear refractive index maximum at 0.01 mol fraction. The production of $[TeO_3]_2$ clusters in the glass matrix is closely related to the rising nonlinear refractive index. The formation of $TeO_3$ clusters is proved from the previous FTIR spectra. The presence of lone pairs in $TeO_3$ structural units leads to an increase in polarizability. As a result, the nonlinear refractive index increased as polarizability increased. Neodymium NPs have a negative impact on the nonlinear refractive index because they cause structural flaws in the glass matrix.

Also, it has been suggested that increasing the concentration of $Nd^{3+}$ ions results in an increasing number of non-bridging oxygen. This is supported by the formation of trigonal pyramidal, $TeO_3$ structural units in the glass system as shown in the FTIR result. The role of non-bridging oxygen contributes to the increasing number of free electrons which are less tightly bound to the nuclear charge.

There is a strong propensity for the free electrons to transition from the valence band to the conduction band. So, the existence of non-bridging oxygen (NBO) leads to the reduction of optical band gap energy. Besides that, the existence of trivalent electrons of neodymium ions affects the structure of the glass system by increasing the number of free electrons. This leads to decreasing number of band gap energy [10]. Also, the decline in $E_{opt}$ is caused by the structural change from $TeO_3$ non-bridging oxygen atoms, the high field strength of neodymium ions, and the high polarizing power of non-bridging oxygen. The value of Urbach energy explains the information about stiffness in the glass matrix.

The compositional dependency of electronic polarizability is displayed in Table 2.1. Increasing neodymium NPs rises, and an upward trend in electronic polarizability can be seen. The deformability of the electron cloud is affected by the number of non-bridging oxygen that exists in the glass system. The non-bridging oxygen (NBO) consists of free electrons from the lone pair which are less tightly bound to the nuclear charge. More than that, the free electrons are easily polarized as the electrical field is penetrated through the medium. The formation of non-bridging oxygen ions is confirmed by the formation of $TeO_3$ and $BO_3$ structural units as shown by FTIR results. The number of non-bridging oxygen (NBO) ions increased as the concentration of dopant increased. The increment number of non-bridging oxygen leads to the raised number of free electrons in the glass system. Hence, the electronic polarizability increased as the concentration of the dopant increased.

Table 2.1 contains the findings about the oxide ion polarizability of tellurite glasses doped with neodymium nanoparticles. The tellurite glasses doped with neodymium NPs have an oxide ion polarizability minimum of 0.03 mol percentage. The tellurite glasses doped with neodymium NPs have an oxide ion polarizability minimum of

0.03 mol percentage. At a specific concentration, zinc oxide has a dual nature that allows it to function as both a modifier and a former glass. Therefore, at 0.03 mol fraction, zinc oxide may behave as former glass, which results in a decrease in the value of oxide ion polarizability. According to Dimitrov and Komatsu [11], the former glass's substantial cationic charge results in a high value for unit field strength. The electron charge density of the nearby oxide ions is significantly impacted by the big value of unit field strength. Therefore, the high concentration of former glass that has a significant cationic charge causes the polarizability of oxide ions to decrease. Furthermore, zinc oxide, $Zn^{4+}$ has a strong unit field strength and a high cationic charge, which results in a reduction in the polarizability of the oxide ions. At 0.05 mol neodymium NPs, oxide ion polarizability reaches its highest value. The strong anionic charge of tellurite, $Te^2$, which has a low unit field strength and a high-refractive index, is the cause of the rising trend in oxide ion polarizability.

In Table 2.1, the information on the optical basicity of the tellurite glass series doped with neodymium NPs is listed. The glass system is more fundamental, as evidenced by the high value of optical basicity in the glass series. The glass system's oxide ions function as a Lewis base, donating electrons to the nearby cations. The acid–base characteristic is significantly influenced by the high degree of donating ability of oxide ions. Additionally, Duffy [12] suggested that the polarization state of the oxide ions in the glass system is connected to the optical basicity. Strong oxide ion donation to the surrounding cations is facilitated by low polarizing cations in the glass system. There is good agreement with optical basicity based on the prior results on oxide ion polarizability.

The nonlinear trend of optical basicity along with concentration was found in this glass series. The nonlinear trend of optical basicity is caused by the structural change in the glass system. Also, the optical basicity values slightly increase with increasing concentration of neodymium NPs. This behavior can be explained by comparing the value of the optical basicity of a single element in the glass system. In the case of glass former and modifier, former oxides should be less basic compared with modifier oxides. When modifier oxide is added to the former oxide, the modification reaction becomes an acid–base reaction in which the modifier oxide ion approaches the former oxide's acidic region in order of decreasing acidities [13].

According to the optical basicity measurements of neodymium NPs-doped tellurite glass, the addition of modifier oxide lowers the former oxide's acidity while raising the basicity of the glass system. The $Nd_2O_3$, $TeO_2$, ZnO, and $B_2O_3$ each have optical basicity values of 1.014, 0.93, 0.82, and 0.425, respectively [14]. As can be seen, borate oxide has a poor optical basicity with a value below 0.75. The optical basicity of borate oxide is increased by the combination of former and modifier oxides. This is caused by the increased optical basicity in zinc oxide (0.82), tellurium oxide (0.93), and neodymium NPs oxide (1.014).

The fundamental function of the acid–base property of glasses is played by tellurite, which also has a considerable impact on glass forming. The ability of oxide ions to give electrons to neighboring cations is increased as optical basicity increases. Table 2.1 lists the results for the tellurite glasses doped with neodymium NPs metallization criteria. The findings demonstrate that when concentration increases, the values of the metallization criterion drop as well. The glass samples are metalized as seen by the decline in the metallization criterion, which causes the band gap to tail off and contract [15].

The band gap of the glass system affects the metallization requirement. The valence and conduction bands have widened, according to the optical band gap's decreasing value. The glass system will become metalized as a result of this process. By contrasting their values, it is possible to determine the direct correlation between the optical band gap and the metallization criterion. The metallization requirement is lowered as the number of optical band gaps rises. Glass samples with low values for the metallization requirement suggest that the glass system forms chemical bonds with a significant ionic contribution. The prior observations of strong optical basicity and high oxide ion polarizability of neodymium NPs-doped tellurite demonstrate the substantial ionicity of the glass samples.

In the year 2019, Azlan et al. [5] measured the electronic polarizability and third-order nonlinearity of Nd3 + -doped borotellurite glass for potential optical fiber. The measured optical nonlinearities such as nonlinear refractive index, nonlinear optical absorption, and third-order susceptibilities were carried out using the Z-scan technique for $\{[(TeO_2)0.70\ (B_2O_3)0.30]0.7\ (ZnO)0.3\}1 - y\ (Nd_2O_3)y$, where $y =$ 0.005, 0.01, 0.02, 0.03, 0.04, and 0.05 glasses. Linear, nonlinear refractive indices, optical band gap energy, $E_{\mathrm{opt}}$ (eV), Fermi energy, $E_F$ (eV), and Urbach energy, $\Delta E$(eV), had been founded [5]. Both $n$ and $E_g$ showed the alteration of tellurite structure by the $Nd^{3+}$. The values of electronic polarizability are enhanced with an increase of $Nd^{3+}$ ions. Figure 2.2 shows the variation of both refractive index and optical band gap energy, $E_g$ with $Nd_2O_3$ mol% and in comparison with silica and chalcogenides. The refractive index increased from 1.76 to 1.86, while the optical band gap energy, $E_{\mathrm{opt}}$, decreased from 3.331 to 3.307 (eV). Besides that, it has been reported that the factors that affect the nonlinearity of the refractive index are as follows:

1. electronic polarizability,
2. electrostriction,
3. molecular orientation,
4. the photorefractive effect, and
5. thermal effect [16].

Moreover, third-order nonlinearity or higher is a component of the nonlinear refractive index. The current work's $Nd^{3+}$ dopant's nonlinear refraction coefficient, ($\eta_o$) ($cm^2/W$), is provided in Table 2.1. It is clear that the Z-scan curves display a peak-to-valley ratio that reflects both self-focusing and positive nonlinear refraction (($\eta_2$) > 0). Self-defocusing and self-focusing are correlated with the nonlinear refractive index's positive or negative sign, depending on the characteristics of the sample. In self-focusing, the incident laser diverges as the sample glass approaches the focal area, increasing irradiance before it reaches the focal point.

The incoming laser is collimated and causes a drop in irradiance as the sample approaches the focal point [17]. Peak-to-valley separation from the laser source is

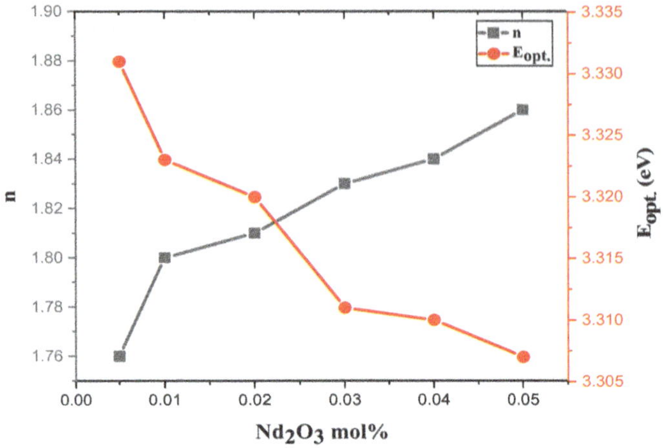


**Fig. 2.2** Variation of both refractive index and optical band gap energy, $E_{opt}$ with $Nd_2O_3$ mol%. Data had been adopted from Ref. [5]

seen to be Roughly 1.7 times Rayleigh length. Thus, it is proven that the third-order reaction is what causes the nonlinear refractive index. Along with dopant concentration, the nonlinear refractive index coefficient for $Nd^{3+}$ also varies nonlinearly. This change can result from the concentration of the dopants. Furthermore, $Nd^{3+}$ has a high nonlinear refractive index. The linear refractive index and nonlinear refractive index have a relationship.

According to earlier linear refractive index measurements, $Nd^{3+}$ dopants have a high linear refractive index value. As a result, it is consistent with the high nonlinear refractive index of $Nd^{3+}$ dopants. Nonlinear refractive index reaches its maximum value at 0.04 and 0.01 mol fraction, respectively. The production of $[TeO_3]_2$—clusters in the glass matrix is closely related to the rising nonlinear refractive index numbers. The earlier FTIR spectra provide evidence for the development of $TeO_3$ clusters. $TeO_3$ structural units become more polarizable when lone pairs are present. As a result, the nonlinear refractive index increased as polarizability increased. The decrease in the nonlinear refractive index at a particular Nd concentration.

Table 2.1 contains the estimated values of the optical band gap ($E_{opt}$), which were determined by extrapolating the absorption coefficient curves. The changes in optical band gap energy are caused by shifts in the valence and conduction bands. The forbidden gap has shrunk, as indicated by the decrease in optical band gap energy. Non-bridging oxygen plays a part in the rise of free electrons that are less closely connected to the nuclear charge. The high valence band to conduction band excitation energy of the unbound electrons result in a decrease in optical band gap energy.

More that, the presence of $TeO_3$ non-bridging oxygen atoms (NBO) contributes to the decrease in optical band gap energy. In Table 2.1, the information on optical basicity for glasses with $Nd^{3+}$ doping is listed. The glass system is more fundamental, as evidenced by the high value of optical basicity in the glass series. The glass system's oxide ions function as a Lewis base, donating electrons to the nearby cations. The acid–base characteristic is significantly influenced by the high degree of donating ability of oxide ions.

Besides the above, it has been suggested [12] that the polarization state of the oxide ions in the glass system is connected to the optical basicity. Strong oxide ion donation to the surrounding cations is facilitated by low-polarizing cations in the glass system. There is good agreement with optical basicity based on the prior results on oxide ion polarizability.

The optical performance of neodymium nanoparticles doped tellurite glasses of $(TeO_2)0.7(B_2O_3)0.3]$ 0.7 $(ZnO)0.31 - x$ $(Nd_2O_3$ NPs) $x$ (where $x = 0.005$, 0.01, 0.02, 0.03, 0.04, and 0.05 mol%) was also reported by Azlina et al. [19]. The highly pure chemical powder of $Nd_2O_3$ (20–30 nm) neodymium(III) oxide nanoparticles (NPs). Due to the addition of ($Nd_2O_3$ NPs), the density increased from 3.719 to 3.936 (g/m$^3$), while the molar volume changed from 31.850 to 32.571 (cm$^3$/mol). These changes are shown in Table 2.1 with $x = 0.005$, 0.01, 0.02, 0.03, 0.04, 0.05 mol%. The creation of non-bridging oxygen was the cause of the increasing density values and neodymium NP concentration [10].

The tendency of the oxide modifier to substitute in the interstices area is increased by the non-bridging oxygen present in the glass system. This pattern will cause the glass network to pack closer together, increasing the glass system's density. The density of the glass samples is impacted by tellurium's and neodymium NPs' different atomic masses. Tellurium (ZTe = 127.6 g·mol$^{-1}$) and neodymium (ZNd = 144.242 g·mol$^{-1}$) have different atomic masses. Since neodymium NPs have an atomic mass larger than tellurium, their concentration causes the density to rise.

The amount of dopants increases along with the molar volume ($V_M$) of neodymium NPs. The relationship between density and molar volume is thought to be inversely proportional. In this instance, the decrease in link length between the atoms may raise the stretching force constant of the bonds in the glass network, which may explain the direct behavior of both density and molar volume [12, 13]. The glass may become denser and more compact as a result.

The optical band gap of $\{(TeO_2)0.7(B_2O_3)0.3]0.7\,(ZnO)0.3\}1 - x\,(Nd_2O_3\ \text{NPs})x$, where $x = 0.005$, 0.01, 0.02, 0.03, 0.04, and 0.05 mol%, is also listed in Table 2.1. It is clear that as neodymium NP concentrations increase, the optical band gap value decreases from 3.209 to 3.178 eV. The valence and conduction bands' movements apart from one another are thought to be responsible for the change in the band gap [15]. The structural alteration in the glass network is closely related to the decrease in optical band gap energy [16]. An increase in the quantity of non-bridging oxygen is caused by the concentration of $Nd^{3+}$ ions. Non-bridging oxygen plays a part in the rise of free electrons that are less closely connected to the nuclear charge. There is a strong propensity for the free electrons to transition from the valence band to the conduction band. Therefore, the energy of the optical band gap decreases in the presence of non-bridging oxygen. In addition, the presence of trivalent neodymium ion electrons alters the structure of the glass system by raising the proportion of free electrons and lowering the band gap energy [17].

Moreover, the structural change from $TeO_3$ non-bridging oxygen atoms, the strong magnetic field of neodymium ions, and the strong polarizing power of non-bridging oxygen all contribute to the drop in $E_{opt}$. The value of Urbach energy explains the information about stiffness in the glass matrix.

Table 2.1 lists the nonlinear refraction coefficient, $\eta_0$ (cm$^2$/W) from 0.734 to 2.98($\times$ 10$^{-12}$), as neodymium NPs-doped tellurite glasses {($TeO_2$)0.7($B_2O_3$)0.3]0.7 (ZnO)0.3}1 $-$ $x$ ($Nd_2O_3$ NPs), where $x$ = mol% Nd2O3 NPs. The nonlinear refractive index coefficient for neodymium NPs is in nonlinear variation, as shown by Table 2.1, coupled with dopant concentration and some prior findings for tellurite glasses [16]. The influence of dopant concentration may be the cause of this difference. The linear refractive index and nonlinear refractive index have a relationship. As previously mentioned, high-refractive index glass materials also have high nonlinear refractive indices [17]. Neodymium NPs have a nonlinear refractive index maximum at 0.01 mol fraction. The production of $[TeO_3]_2$—clusters in the glass matrix is closely related to the rising nonlinear refractive index numbers.

The earlier FTIR spectra are evidence that $TeO_3$ clusters had formed [18]. $TeO_3$ structural units become more polarizable when lone pairs are present. As a result, the nonlinear refractive index increased as polarizability increased. Neodymium NPs have a negative impact on the nonlinear refractive index because they cause structural flaws in the glass matrix.

## References


1. Simmons, C.J., and O. El-Bayoumim. 1993. Experimental techniques of glass science. In: *Amer, Ceramic Society*.
2. Boling, N., A. Glass, and A. Owyoung. 1978. Empirical relationships for predicting nonlinear refractive index changes in optical solids. *IEEE Journal of Quantum Electronics* 14 (8): 601–608.
3. Duffy, J. 1986. Chemical bonding in the oxides of the elements: A new appraisal. *Journal of Solid State Chemistry* 62 (2): 145–157.
4. Dimitrov, V., and S. Sakka. 1996. Linear and nonlinear optical properties of simple oxides. II. *Journal of Applied Physics* 79 (3): 1741–1745.
5. Azlan, M., et al. 2019. Electronic polarizability and third-order nonlinearity of $Nd^{3+}$ doped borotellurite glass for potential optical fiber. *Materials Chemistry and Physics* 236: 121812.
6. Shanmugavelu, B., V. Venkatramu, and V.R.K. Kumar, Optical properties of $Nd^{3+}$ doped bismuth zinc borate glasses. *Spectrochimica Acta Part A: Molecular and Biomolecular Spectroscopy* 122: 422–427.
7. Umar, S., et al. 2023. Optical basicity, polarizability and spectroscopic investigations of CuO doped $TeO_2$–$B_2O_3$ glass system. *Materials Chemistry and Physics* 2023: 127309.
8. Azlan, M., et al. 2018. Linear and nonlinear optical efficiency of novel neodymium nanoparticles doped tellurite glass for advanced laser glass. *Educational JSMT* 5 (2): 47–66.
9. Chen, Y., et al. 2008. A study of nonlinear optical properties in $Bi_2O_3$–$WO_3$–$TeO_2$ glasses. *Journal of Non-Crystalline Solids* 354 (29): 3468–3472.
10. Maheshvaran, K., P. Veeran, and K. Marimuthu. 2013. Structural and optical studies on $Eu^{3+}$ doped boro-tellurite glasses. *Solid State Sciences* 17: 54–62.
11. Dimitrov, V., and T. Komatsu. 2010. An interpretation of optical properties of oxides and oxide glasses in terms of the electronic ion polarizability and average single bond strength. *Journal of University in Chemical Technology Metallaurgy* 45 (3): 219–250.

12. Duffy, J.A. 1989. Optical basicity of titanium(IV) oxide and zirconium(IV) oxide. *Journal of the American Ceramic Society* 72 (10): 2012–2013.
13. Sidek, H., et al. 2004. Formation and elastic behavior of lead-magnesium chlorophosphate glasses. *Turkish Journal of Physics* 28 (1): 65–72.
14. Dimitrov, V., and T. Komatsu. 2005. Classification of oxide glasses: A polarizability approach. *Journal of Solid State Chemistry* 178 (3): 831–846.
15. Sun, J., et al. 2011. Glass formation and properties of Ge–Te–$BiI_3$ far infrared transmitting chalcohalide glasses. *Spectrochimica Acta Part A: Molecular and Biomolecular Spectroscopy* 79 (5): 904–908.
16. Iranizad, E.S., Z. Dehghani, and M. Nadafan. 2014. Nonlinear optical properties of nematic liquid crystal doped with different compositional percentage of synthesis of $Fe_3O_4$ nanoparticles. *Journal of Molecular Liquids* 190: 6–9.
17. Mojdehi, M.S., et al. 2013. Nonlinear optical characterization of phosphate glasses based on ZnO using the Z-scan technique. *Chinese Physics B* 22 (11): 117802.
18. Azlan, M., and M. Halimah. 2018. Role of $Nd^{3+}$ nanoparticles on enhanced optical efficiency in borotellurite glass for optical fiber. *Results in Physics* 11: 58–64.
19. Azlina, Y., et al. 2020. Optical performance of neodymium nanoparticles doped tellurite glasses. *Physica B: Condensed Matter* 577: 411784.
20. Vasconcelos, H. 2022. *Optical Nonlinearities in Glasses, Open Access Peer-Reviewed Chapter*. https://doi.org/10.5772/intechopen.101774.

# Chapter 3
# Ultraviolet Absorption Spectra of Some Tellurite Glasses

The present chapter deals with the interaction of the UV spectra with some tellurite glasses, glass ceramics, and ceramics. Also, it contains five sections about measurements of UV spectra, theoretical background, experimental data with the impact of rare-earth NPs, tellurite glass ceramics doped by a transition metal, and quantum dot (QD)-embedded tellurite glasses.

## 3.1 Measurements of UV Spectra

A useful instrument for examining the optical absorption of the materials at hand is UV–Vis spectroscopy. Additionally, examination of the optical absorption reveals the glass samples' optical band gap and Urbach energy. To determine the optical absorption spectra at room temperature and in the range of 190–1100 nm of wavelength, a UV–Vis spectrometer (Shimadzu, model: UV 1650) was employed. Figure 3.1 illustrates the UV–visible absorption principle. The fundamental idea behind UV–visible spectroscopy is based on how various chemical compounds absorb light at different wavelengths starting at -nm. The sample holder was filled with the thin sample of glass (< 2 mm in thickness). For the characterization process, a Xenon light flash was used as the light source. The computer system connected to the equipment recorded the optical absorption spectra. A wavelength precision of 0.1 nm is probably sufficient, and the peaks must be stated with a tolerance of 1 nm.

 
R. El-Mallawany, *The Physics of Advanced Optical Materials: Tellurite Glasses*,
https://doi.org/10.1007/978-3-031-45245-1_3

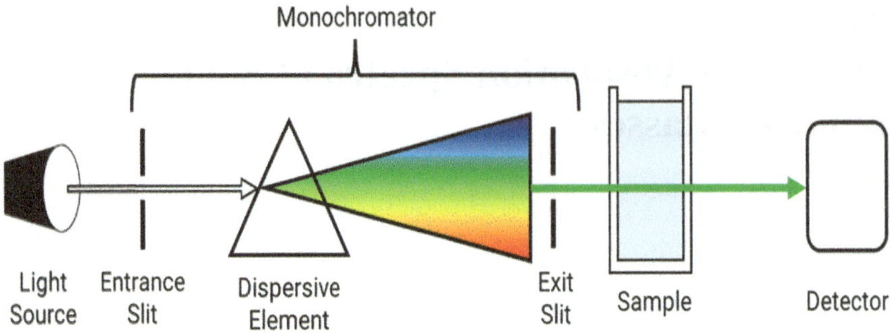


**Fig. 3.1** Principle of UV–visible absorption

## 3.2 Theoretical Background

The absorption coefficients $\alpha$ ($\nu$) for all of the glasses analyzed were determined directly from the absorbance measurements ($A$) using the following equation:

$$\alpha(\nu) = 2.303\left(\frac{A}{d}\right) \quad (3.1)$$

where $A$ = absorbance in (a.u), and $d$ = the sample thickness in cm. The ($\lambda_{edge}$) is found that the value of the wavelength edge. The optical energy gap, $E$ (eV), was calculated using Tauc's model [1], and it was developed by Losche et al. [2] written as:

$$\alpha(\nu)h\nu = C(h\nu - E^{\mathrm{Tauc's}})^k \quad (3.2)$$

where $C$ is constant subject to the transition probability, $k$ = 0.5 (direct) or 2 (indirect) allowed transitions, respectively. Then the changes of $(\alpha h\nu)^{1/2}$ and $(\alpha h\nu)^2$ with the photon energy for indirect and direct allowed transition, respectively. Values of the optical energy gap ($E^{\mathrm{Tauc's}}$) were calculated from the extrapolating linear part of each spectrum. Table 3.1 collected optical energy gap ($E^{\mathrm{Tauc's}}$). The absorption spectra fitting (ASF) method is another way applied to evaluate the optical energy gap. Souri and Shomalian [3], Souri and Tahan [4], Escobar-Alarcon et al. [5], and Ali et al. [6] showed that the $\alpha(\nu)$ can be converted to:

$$\alpha(\lambda) = C(hc)^{k-1}\lambda\left(\lambda^{-1} - \lambda^{-1}_{\text{cut - off}}\right)^k \quad (3.3)$$

**Table 3.1** Optical energy gap of some ordinary tellurite glasses and glasses containing nanoparticles

| Glass mol% | References | Calculated refracted index | Optical energy band (eV) |
|---|---|---|---|
| Ordinary glass $\{[(TeO_2)0.70\ (B_2O_3)0.30]0.70\ (ZnO)0.30\}_{1-y}\ (Er_2O_3)$ | [14] | | |
| 0.005 | | 1.716 | 3.87 |
| 0.010 | | 1.718 | 3.85 |
| 0.020 | | 1.721 | 3.85 |
| 0.030 | | 1.730 | 3.85 |
| 0.040 | | 1.732 | 3.85 |
| 0.050 | | 1.740 | 3.85 |
| Glass with nanoparticles, $y(Er_2O_3$ nanoparticles) | | | |
| 0.005 | | 1.774 | 3.35 |
| 0.010 | | 1.818 | 3.00 |
| 0.020 | | 1.830 | 3.30 |
| 0.030 | | 1.855 | 3.25 |
| 0.040 | | 1.878 | 3.45 |
| 0.050 | | 1.923 | 3.40 |
| $\{(TeO_2)0.7(B_2O_3)0.3]0.7\ (ZnO)0.3\}_{1-x}\ (Nd_2O_3$ NPs) $x$ (where $x = 0.005, 0.01, 0.02, 0.03, 0.04$ and $0.05$) | [17] | Urbach energy, ΔE (eV) | |
| 0.005 | | 0.316 | 3.209 |
| 0.01 | | 0.312 | 3.201 |
| 0.02 | | 0.314 | 3.198 |
| 0.03 | | 0.316 | 3.193 |
| 0.04 | | 0.318 | 3.179 |
| 0.05 | | 0.323 | 3.178 |
| $\{[(TeO_2)0.70\ (B_2O_3)0.30]0.7\ (ZnO)0.3\}_{1-y}\ (Sm_2O_3$ (MPs/NPs))$y$ with $y = 0.005, 0.01, 0.02, 0.03, 0.04$ and 0.05 mol%. ZBTSm-MPs | [18] | Refractive index | Indirect<br>Direct (eV) |
| 0.000 | | 1.871 | 2.780<br>2.845 |
| 0.005 | | 1.893 | 2.528<br>2.804 |
| 0.010 | | 1.974 | 2.898<br>2.970 |

(continued)

**Table 3.1** (continued)

| Glass mol% | References | Calculated refracted index | Optical energy band (eV) |
|---|---|---|---|
| 0.020 | | 1.873 | 2.921<br>2.970 |
| 0.030 | | 1.968 | 2.851<br>2.951 |
| 0.040 | | 2.030 | 2.782<br>2.920 |
| 0.050 | | 2.005 | 2.845<br>2.943 |
| ZBTSm-NPs | [18] | | |
| 0.005 | | 2.056 | 2.956<br>3.006 |
| 0.010 | | 2.011 | 2.985<br>3.021 |
| 0.020 | | 2.007 | 3.002<br>3.021 |
| 0.030 | | 2.045 | 2.991<br>3.021 |
| 0.040 | | 2.050 | 2.999<br>3.019 |
| 0.050 | | 2.051 | 2.980<br>3.009 |

The cut-off wavelength is represented by $\lambda_{\text{cut - off}}$, $c =$ the velocity of light, and $h$ = Planck's constant. Equation (3.3) can be simplified to:

$$A(\lambda) = D\lambda\left(\lambda^{-1} - \lambda_{\text{cut - off}}^{-1}\right)^{k} \tag{3.4}$$

where $D = [C(hc)p - 1\ d/2.303]$. The band gap in the ASF method was estimated directly from $\lambda_{\text{cut - off}}$ as:

$$E^{\text{ASF}} = hc\left(\lambda_{\text{cut - off}}^{-1}\right) = 1239.83\left(\lambda_{\text{cut - off}}^{-1}\right) \tag{3.5}$$

By extrapolating the linear part of these plots at the values of $(A/\lambda)1/2$ and $(A/\lambda)2 = 0$, the $\lambda_{\text{cut - off}}$ value for each plot was determined. Substituting the values of $\lambda_{\text{cut - off}}$ Eq. (3.5), the corresponding ($E^{\text{ASF}}$) will be calculated. The $E^{\text{ASF}}$ values for all studied glasses were listed in Table 3.1. Values of $E^{\text{ASF}}$ were in agreement with that obtained on Tauc's method. In addition, the values of the energy gap for the indirect case were lower than that in the direct case. The refractive index ($n$) of glasses was calculated by using the next relation [7–9].

$$\left(\frac{n^2-1}{n^2+2}\right)=1-\sqrt{\frac{E_{\text{Opt.}}^{\text{AsF}}}{20}} \tag{3.6}$$

The molar refraction ($R_m$) of material mainly depends on ($V_m$) and ($n$), and it can be given by the next equation:

$$R_m=\left(\frac{n^2-1}{n^2+2}\right)V_m \tag{3.7}$$

The electronic polarizability ($\alpha_m$) of the investigated glasses can be expressed by the next equation:

$$\alpha_m=\frac{R_m}{2.52} \tag{3.8}$$

Moreover, reflection loss ($R_L$) and optical transmission ($T$) can be calculated by using the next equations, respectively,

$$R_L=\left(\frac{n-1}{n+1}\right)^2 \tag{3.9}$$

$$T=\frac{2n}{n^2+1} \tag{3.10}$$

The metallization property is the final optical parameter, which helps to determine the material's existence (metallic or nonmetallic) [10]. The metallization ($M$) was calculated by using the next equations [11, 12], respectively:

$$M(n)=1-\left(\frac{n^2-1}{n^2+2}\right) \tag{3.11}$$

$$M\left(E_{\text{Opt.}}^{\text{ASF}}\right)=\left(\frac{E_{\text{Opt.}}^{\text{ASF}}}{20}\right)^{1/2} \tag{3.12}$$

Urbach [13] rule was written in the form

$$\alpha(\omega)=B\,\exp\left(\frac{\hbar\omega}{\Delta E}\right) \tag{3.13}$$

where ($B$) = $a$ constant, and ($\Delta E$) = the width of the band tails of the localized states.

## 3.3 Some Experimental Data, Fermi Energy, Optical Band Gap, and Impact of R.E.NPs

As noted in Sects. 3.2–3.5, optical characteristics between two glass series (*a*—ordinary glass, and *b*—glass containing nanoparticles) have been observed [14]. Figure 3.2a represents the UV–visible absorption spectra and extinction coefficient versus wavelength for $\{[(TeO_2)_{0.70}(B_2O_3)_{0.30}]_{0.7}(ZnO)_{0.3}\}_{1-y}(Nd_2O_3)_y$ where, $y =$ 0.005, 0.01, 0.02, 0.03, 0.04, and 0.05 glasses [15]. Also, $TeO_2$–$B_2O_3$ glasses' spectroscopic characteristics, polarizability, and optical basicity have all been described [15]. The refractive index (2.318–2.378), the metallization criteria (0.4069–0.3919), optical basicity (0.7995–0.8057), and electronic polarizability (0.2358–0.2410) were among the physical and optical parameters studied. The third-order nonlinear dielectric susceptibility (1.466 1012–1.882 1012 esu) and linear dielectric susceptibility (0.3480–0.3704 esu) have been established. The applications of the glasses in fiber optics and laser technology could be quite promising.

The extinction coefficient spectra of the investigated glass were depicted in Fig. 3.2b [15]. The glass system's extinction coefficient spectra exhibit a sharp increase near the ultraviolet (UV) region, indicating a significant value in the UV area. It was discovered that as $Nd^{3+}$ levels rose, the extinction coefficient dropped. The following values of the Fermi–Dirac distribution function can be used to compute the Fermi energy $E_F$ (eV) of the glass system: 3.686, 3.611, 3.491, 3.441, 3.301, and 3.270 (eV). The table shows that as $Nd^{3+}$ levels rise, the Fermi energy $E_F$ (eV) value decreases. The improvement of semiconducting characteristics is indicated by the decrease in Fermi energy $E_F$ (eV). More than that, it is discovered that the Fermi energy ($E_F$) value (eV) is lower than the thermal energy at room temperature (kBT 25 meV). Furthermore, the glass system has a propensity toward semiconducting behavior. A study on spectroscopic observations of a CuO-doped $TeO_2$–$B_2O_3$ glass system was published in 2023 [16]. Table 3.1 contains the optical energy gaps of various plain tellurite glasses and nanoparticle-containing tellurite glasses.

Figure 3.3 shows that both the $E_{opt}$ (eV) and refractive index of some tellurite glasses are different from ordinary glass, glass doped with MPs, and glass doped with NPs. Previously, it has been reported [15] that band gap increases as the size of particles decreases. Therefore, ($E_{opt}$) has been reported to be size-dependent. So, there is an increase in the band gap of the semiconductor with a reduction in the particle size. Besides that, Er (NPs) are made up of only finite number of particles. This will appear in the number of overlapping orbitals decreasing and the width of the bands starting to narrow. The reduction in the optical band gap energy indicates that the forbidden gap is decreased. Furthermore, the structural change in the glass network leads to the shifts of the forbidden gap which affect the optical band gap energy. The role of non-bridging oxygen was to contribute to the increasing number of free electrons which are less tightly bound to the nuclear charge. While free electrons have a high ability to excite from the valence band to the conduction band which in turn leads to the reduction of optical band gap energy. Besides the above, the inclusion of $Nd^{3+}$ ions in the glass system increases the number of weaker bonds

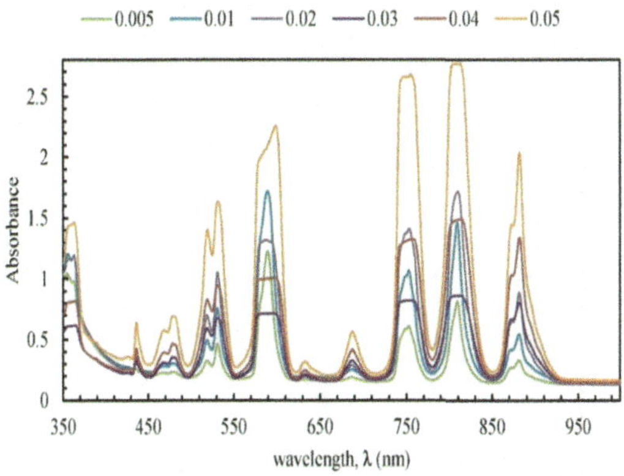


(a)

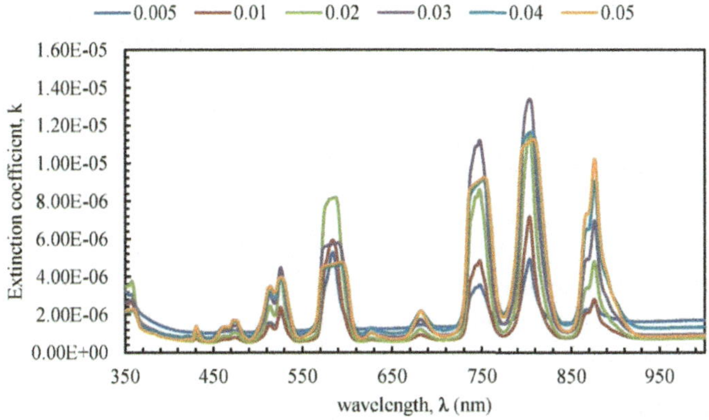


(b)

**Fig. 3.2** **a, b** UV–visible absorbance and extinction coefficient versus wavelength for $\{[(TeO_2)_{0.70} (B_2O_3)_{0.30}]_{0.7} (ZnO)_{0.3}\}_{1-y} (Nd_2O_3)_y$ glasses [15]. With permission

which reduces the number of optical band gap energy. Also, the decline in $E_{opt}$ energy is caused by the presence of $TeO_3$ non-bridging oxygen atoms, the high field strength of $Nd^{3+}$, and the high polarizing power of NBO.

The precise values of the direct and indirect optical band gaps are listed in Table 3.1. The ZBTSm-NPs had a wider optical band gap than ZBTSm-MPs [18]. The optical band gap is widened by the tiny particles' restriction on valence electron mobility. The optical characteristics of materials, according to Gupta and Ramrakhiani [19], are greatly influenced by particle size. Previous studies demonstrated that the $E_{opt}$ is size dependent, and that a semiconductor's band gap increases when particle size decreases [20, 21]. The tiny samarium oxide particles increase the space between the valence and conduction bands by reducing the number of overlapping orbitals.

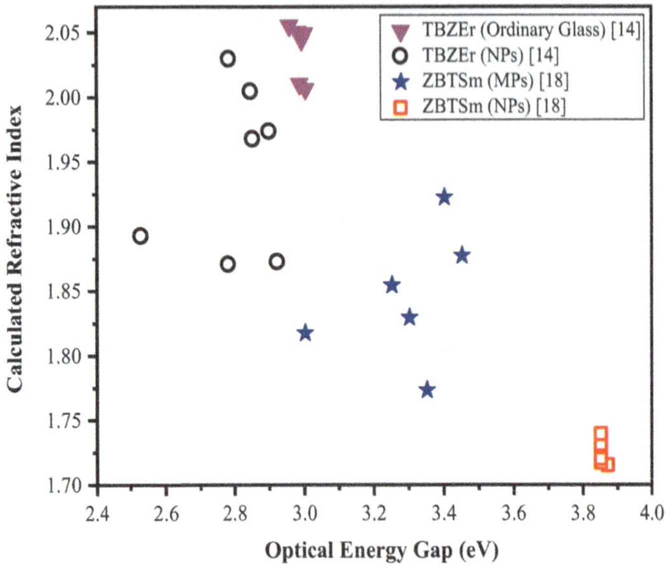


**Fig. 3.3** $E_{opt}$ (eV) and calculated refractive index of some tellurite glasses (Compositions are as in Table 3.1)

According to a conclusion [18], the following differences in structural, physical, and optical properties between ZBTSm-MPs and ZBTSm-NPs glasses were significant:

(1) The 23.53 nm was found for the average size of nanoparticles in ZBTSm-NPs,
(2) Because nanoscale particles cause the glass structure to become more compact, ZBTSm-NPs glasses were shown to be less dense than ZBTSm-MPs glasses,
(3) The difference in density between ZBTSm-NPs and ZBTSm-MPs glasses was shown to increase the refractive index of the former,
(4) ZBTSm-MPs glasses had absorption peaks that were twice as high as ZBTSm-NPs glasses, which is consistent with the confinement of electrons in nanoscale particles,
(5) The $E_{opt}$ of (ZBTSm-NPs) glasses was found greater than (ZBTSm-MPs) glasses which is mainly due to the widening of the forbidden gap with nanoscale particles,
(6) Nonlinear trend of polarizability was found in both sets of glasses due to the role of zinc oxide in the tellurite glass system.

## 3.4 Transition Metal Doped Tellurite Glass-Ceramics

Depending on the outer shell electric configuration of their various oxidation states, transition metal ions (TM) like Cr, Mn, Cu, and Co can exhibit good luminescence in the visible and NIR wavelength areas [22]. The emission spectra of TMs are far more sensitive to the ligand field than those of RE ions. Since the immediate environment might alter as a result of crystallization, this can have a significant effect on how bright they are. Nanostructured glass ceramics may enable a longer lasting, wideband adjustable emission.

In addition, multivalent TMs can exist in $TM^{1+}$ through $TM^{4+}$ states and, depending on the host environment, can adopt either tetrahedral or octahedral coordination. Green and red emissions from $Mn^{2+}$ and $Cr^{3+}$, respectively, are well-known properties that can efficiently sensitize RE ions [23, 24]. The $Ni^{2+}$-doped glass ceramics have been reported to display ultra-broad tunable emission in the NIR region (1100–1800 nm) [25]. There are not many reports on $Cr^{4+}$ and $V^{3+}$'s effective broadband emission in the NIR range. $Al_2O_3$:$Cr^{3+}$ nanocrystals made of a tellurite glass composite were synthesized by Han et al. [26], who then suggested that they would work well as a spectrum converter in silicon solar cells.

In the $Al_2O_3$ lattice, $Cr^{3+}$ ions act as a broad UV–Vis (320–670 nm) spectral acceptor, which effectively converts into strong red emission (696 nm) and easily overcomes the so-called spectral mismatch in the silicon solar cell, according to authors' explanations. As mentioned in other glasses and glass ceramics [27], TM ions can also facilitate a quantum-cutting process and contribute to raising the system's overall quantum efficiency.

## 3.5 Quantum Dot-Embedded Tellurite Glass

Aside from their bulk, quantum well, or quantum wire counterparts, semiconductor quantum dots (QDs) are unique as an optical gain medium due to their three-dimensional quantum confinement effect, high transition oscillator strength, the density of state that resembles a delta function, and temperature-independent low threshold optical transition [28]. Due to their highly intriguing emission features, which result from exciting recombination at cationic vacancies (PbX6 4 octahedron), all inorganic halide-based perovskites ($CsPbX_3$, $X = $ Cl, Br, I) QDs have recently been extensive research [22]. The QDs' Bohr diameters of up to 12 nm contribute to their capacity to provide band gap tunability over the whole visible spectrum (410–700 nm). These perovskites are said to exhibit photoluminescence with a high quantum yield of 70–90% upon UV excitation, making them suitable for efficient solid-state lighting, upconverted random lasing, solar cells, photodetectors, liquid–crystal displays, and other optoelectrical applications [29] (Fig. 3.4).

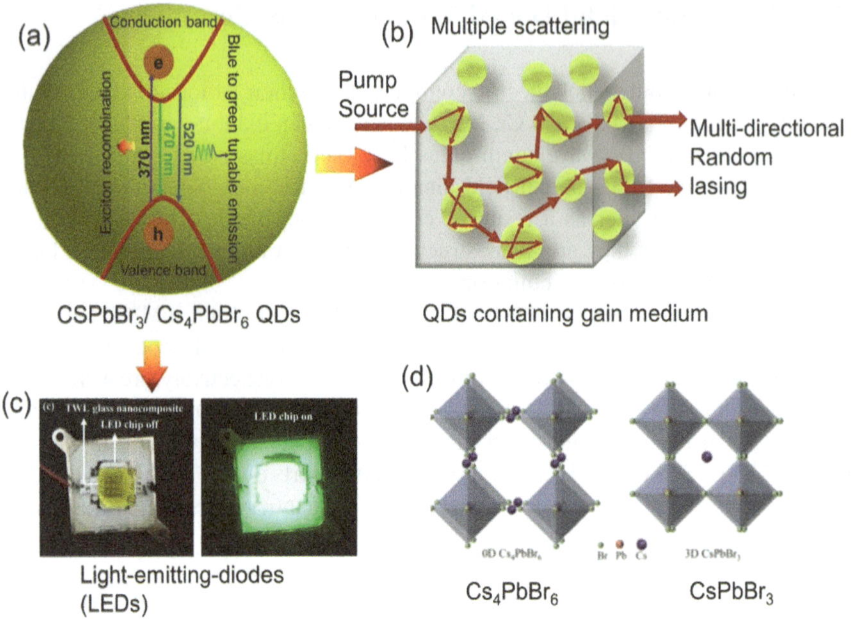


**Fig. 3.4** **a** Probable pictorial exciton recombination process assisted emission mechanism within $CsPbBr_3$ and $Cs_4PbBr_6$ QDs, **b** Random lasing process through multiple scattering in a QDs embedded tellurite glass composite, **c** Demonstration of green light emission of tellurite glass @ QDs nanocomposite in combination with blue light-emitting LED chip [30], **d** A probable structural comparison between these two $CsPbBr_3$ and $Cs_4PbBr_6$ QDs [31]. With permission

# References


1. Tauc, J. 2012. *Amorphous and liquid semiconductors*. New York: Springer.
2. Lösche, A., N.F. Mott, and E.A. Davis. 1972. *Electronic processes in non-crystalline materials clarendon-press*. Amsterdam: Wiley Online Library.
3. Souri, D., and K. Shomalian. 2009. Band gap determination by absorption spectrum fitting method (ASF) and structural properties of different compositions of (60–x) $V_2O_5$–40$TeO_2$–x$Sb_2O_3$ glasses. *Journal of Non-Crystalline Solids* 355 (31–33): 1597–1601.
4. Souri, D., and Z.E. Tahan. 2015. A new method for the determination of optical band gap and the nature of optical transitions in semiconductors. *Applied Physics B* 119 (2): 273–279.
5. Escobar-Alarcón, L., et al. 2007. An alternative procedure for the determination of the optical band gap and thickness of amorphous carbon nitride thin films. *Applied Surface Science* 254 (1): 412–415.
6. Ali, A., et al. 2017. FTIR and UV spectra of pentaternary borate glasses. *Measurement* 105: 72–77.
7. Dimitrov, V., and S. Sakka. 1996. Electronic oxide polarizability and optical basicity of simple oxides. *I. Journal of Applied Physics* 79 (3): 1736–1740.
8. 小野周. 1961. EA Moelwyn-Hughes: Physical chemistry, pergamon press, 1961, 1333 ページ. *83 S. 日本物理学会誌* 16(12): 798.
9. El-Mallawany, R. 1992. The optical properties of tellurite glasses. *Journal of Applied Physics* 72 (5): 1774–1777.
10. Rawson, H. 1980. *Properties and applications of glass*, vol. 3. Amsterdam: Elsevier.

11. Dimitrov, V., and S. Sakka. 1996. Linear and nonlinear optical properties of simple oxides. II. *Journal of Applied Physics* 79 (3): 1741–1745.
12. Vithal, M., et al. 1997. Optical and electrical properties of PbO–$TiO_2$, PbO–$TeO_2$, and PbO–CdO glass systems. *Journal of Applied Physics* 81 (12): 7922–7926.
13. Urbach, F. 1953. The long-wavelength edge of photographic sensitivity and of the electronic absorption of solids. *Physical Review* 92 (5): 1324.
14. Azlan, M., et al. 2017. Optical properties of zinc borotellurite glass system doped with erbium and erbium nanoparticles for photonic applications. *Journal of Materials Science: Materials in Electronics* 28: 4318–4327.
15. Azlan, M., et al. 2019. Electronic polarizability and third-order nonlinearity of $Nd^{3+}$ doped borotellurite glass for potential optical fiber. *Materials Chemistry and Physics* 236: 121812.
16. Umar, S., et al. 2023. Optical basicity, polarizability and spectroscopic investigations of CuO doped $TeO_2$–$B_2O_3$ glass system. *Materials Chemistry and Physics* 14: 127309.
17. Azlan, M., et al. 2018. Linear and nonlinear optical efficiency of novel neodymium nanoparticles doped tellurite glass for advanced laser glass. *Educational JSMT* 5 (2): 47–66.
18. Azlan, M., et al. 2021. Comprehensive comparison on optical properties of samarium oxide (micro/nano) particles doped tellurite glass for optoelectronics applications. *Journal of Materials Science: Materials in Electronics* 32 (11): 14174–14185.
19. Gupta, P., and M. Ramrakhiani. 2009. Influence of the particle size on the optical properties of CdSe nanoparticles. *The Open Nanoscience Journal* 3 (1): 53–59.
20. Kassab, L.R.P., et al. 2011. Effects of gold nanoparticles in the green and red emissions of $TeO_2$–PbO–$GeO_2$ glasses doped with $Er^{3+}$–$Yb^{3+}$. *Optical Materials* 33 (12): 1948–1951.
21. Qi, Y., et al. 2014. Silver nanoparticles enhanced 1.53 μm band fluorescence of $Er^{3+}$/$Yb^{3+}$ codoped tellurite glasses. *Journal of Luminescence* 153: 401–407.
22. Patra, P., and K. Annapurna. 2022. Transparent tellurite glass-ceramics for photonics applications: A comprehensive review on crystalline phases and crystallization mechanisms. *Progress in Materials Science* 125: 100890.
23. Dan, H.K., et al. 2020. Energy transfer and spectroscopic properties of $Cr^{3+}$/$Yb^{3+}$ co-doped $TeO_2$–ZnO–$La_2O_3$ tellurite glasses under different wavelength excitation lights. *Optical Materials* 100: 109662.
24. Ghosh, D., et al. 2017. Tunable white light generation from $Ce^{3+}$–$Tb^{3+}$–$Mn^{2+}$ doped metaphosphate glass for LED and solar cell applications. *Journal of Luminescence* 183: 143–149.
25. Lin, C., et al. 2016. Oxyfluoride glass-ceramics for transition metal ion based photonics: Broadband near-IR luminescence of nickel ion dopant and nanocrystallization mechanism. *The Journal of Physical Chemistry C* 120 (8): 4556–4563.
26. Han, B., et al. 2015. $Al_2O_3$: $Cr^{3+}$/tellurite glass composites: An efficient light converter for silicon solar cell. *Ceramics International* 41 (9): 12267–12272.
27. Jiménez, J.A. 2015. UV emission of $Gd^{3+}$ in the presence of $Cu^{2+}$: Towards luminescence quenching through quantum cutting? *ChemPhysChem* 16 (8): 1683–1686.
28. Wang, Y., et al. 2015. All-inorganic colloidal perovskite quantum dots: A new class of lasing materials with favorable characteristics. *Advanced Materials* 27 (44): 7101–7108.
29. Song, J., et al. 2015. Quantum dot light-emitting diodes based on inorganic perovskite cesium lead halides ($CsPbX_3$). *Advanced in Materials* 27: 7162–7167.
30. Erol, E., et al. 2020. Size-controlled emission of long-time durable $CsPbBr_3$ perovskite quantum dots embedded tellurite glass nanocomposites. *Chemical Engineering Journal* 401: 126053.
31. Duan, Y., et al. 2021. Blue-green tunable luminescence mechanism of a novel $Tb^{3+}$ doped $Cs_4PbBr_6$ quantum dots tellurite glass. *Journal of Luminescence* 231: 117799.

# Chapter 4
# Some Physical Characteristics of Tellurite Glasses and Tellurite Glass Ceramics

Due to its greater optical nonlinearity and superior optical quality, scientists have recently increased their focus on glassy materials. It is commonly believed that tellurium oxide is a conditional glass maker and that it can only form glasses in the presence of some transitional metal oxides, rare-earth oxides, alkalis, and some other modifiers in the glass lattice. A conditional glass former, $TeO_2$, can only produce glass when combined with other glass formers or modifiers like alkali, alkaline earth, or transition metal oxide (TMO) and has unique physical properties different from borate or phosphate glasses. Recently tellurite glasses have been achieved for the operation of tuneable continuous-wave laser on tellurite glass doped with $RE^{3+}$ ions. Tellurite glasses are different from other glasses in terms of their physical characteristics, and they are used in energy conversion technology, temperature sensing, laser devices, and energy harvesting. These glasses have been termed "smart materials" because of their distinctive physical characteristics.

The present chapter will focus on some optical characteristics of tellurite glasses with the physical interpretations as follows:

1. Thermoluminescence dosimetry,
2. Luminescence properties and spectra,
3. Emission and absorption analysis,
4. White light source and optical thermometry,
5. Impact of rare-earth nanoparticles on up-conversion, down-conversion, and thermal properties,
6. Tellurite glasses, tellurite glass ceramics, and tellurite ceramics for different purposes.

 
R. El-Mallawany, *The Physics of Advanced Optical Materials: Tellurite Glasses*,
https://doi.org/10.1007/978-3-031-45245-1_4

## 4.1 Thermoluminescence Dosimetry (TLD)

In the early 1990s [1], thermoluminescence (TL) of $TeO_2$-$P_2O_5$ glasses doped with $CeO_2$, $Sm_2O_3$, and $Yb_2O_3$ has been measured as shown in Fig. 4.1. Depending on the kind of rare-earth oxide and the gamma radiation dose for tellurite-phosphate glass doped with 0.001 weight percent, the glow curve peaked at roughly 485–535 K. In the dose range of 0–200 μGy, it was discovered that both the peak height and the area under TL glow curves were linear. The gamma-ray source $^{60}Co$ was used to irradiate the surface. The following reactions may be used to mimic the probable TL mechanisms occurring under gamma-ray irradiation:

$$\mathrm{RE}^{3+} + (\text{gamma - ray}) = \mathrm{RE}^{2+} + h^{+} \tag{4.1}$$

$$\mathrm{RE}^{3+} + (\text{gamma - ray}) = \mathrm{RE}^{3+} + h^{+} + e^{-} \tag{4.2}$$

where $h^+$ and $e^-$ represent holes and electrons, respectively. In Eq. (4.1), as the material is heated, a hole that was trapped at a lattice site during irradiation recombines with $RE^{2+}$ to produce TL. In Eq. (4.2), during irradiation, both electrons and holes are trapped in the lattice; after heating up during TL, the electron and hole recombine at the $RE^{3+}$ site.

After that, thermoluminescence for rare earths-doped tellurite glasses has been reported by Elkholy [2]. Moreover, in 2012, thermoluminescence (TL) properties of the $80(TeO_2)$–$5(TiO_2)$–$(15 - x)(WO_3)$–$(x)A_nO_m$, where $A_nO_m = Nb_2O_5$, $Nd_2O_3$, $Er_2O_3$, and $x$ mol% have been identified [3]. The TL main peak of dosimetry for the glasses was investigated for $^{60}Co$ gamma rays as shown in Fig. 4.2a–d. It has been determined that this material can be used in a phantom for testing treatment plans for dose administration soon since the therapeutic range (0.5–2 Gy) of the TL response versus the administered dose without reaching saturation level [3] is present.

## 4.2 Luminescence Properties and Spectra

In 2004, the effects of the $Er^{3+}$ ions concentration on the decay time, spectral breadth, and weighted spectral peak in tellurite glasses have been reported [4] as shown in Fig. 4.3a–d.

Results [4] revealed that the local crystal field generated in these new glasses [5, 6] helps to broadband $Er^{3+}$ emission at 1550 nm as follows:

| Glass composition (mol%) | Weighted peak (nm) | FWHM (nm) | Decay time (ms) |
|---|---|---|---|
| $80TeO_2$–$5TiO_2$–$14.99WO_3$–$0.01Er_2O_3$ | 1541 | 93 | $4.0 \pm 0.6$ |
| $80TeO_2$–$5TiO_2$–$10WO_3$–$5Er_2O_3$ | 1554 | 120 | $1.00 \pm 0.01$ |

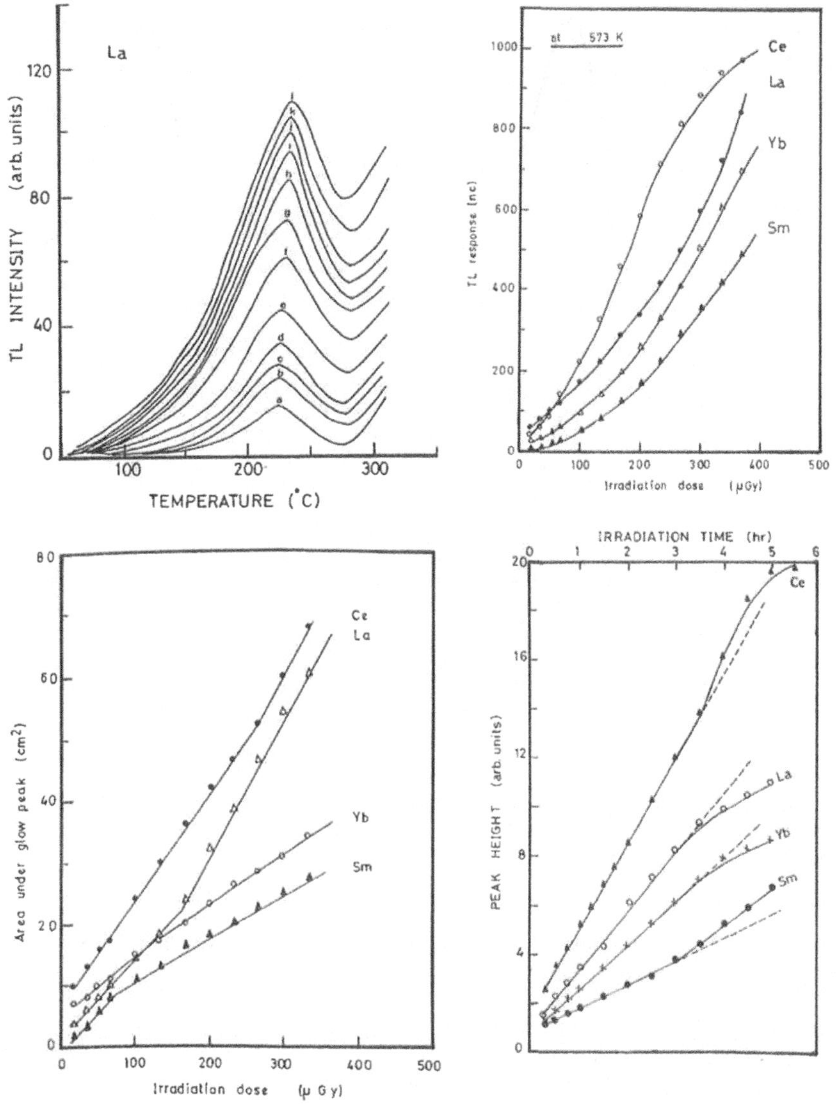


**Fig. 4.1** Glow curves of gamma-irradiated $TeO_2$–$P_2O_5$ glasses doped with $CeO_2$, $Sm_2O_3$, $Yb_2O_3$, and $La_2O_3$ [1] with permission

Then in 2009, both absorption and emission analyses of $RE^{3+}$ ($Sm^{3+}$ and $Dy^{3+}$): lithium borotellurite glasses have been reported [7]. After that, in 2011, luminescence spectra and optical properties of $TeO_2$-$WO_3$–$Li_2O$ glasses doped with Nd, Sm, and Er rare-earth ions of tellurite glasses were measured at room temperature by using a

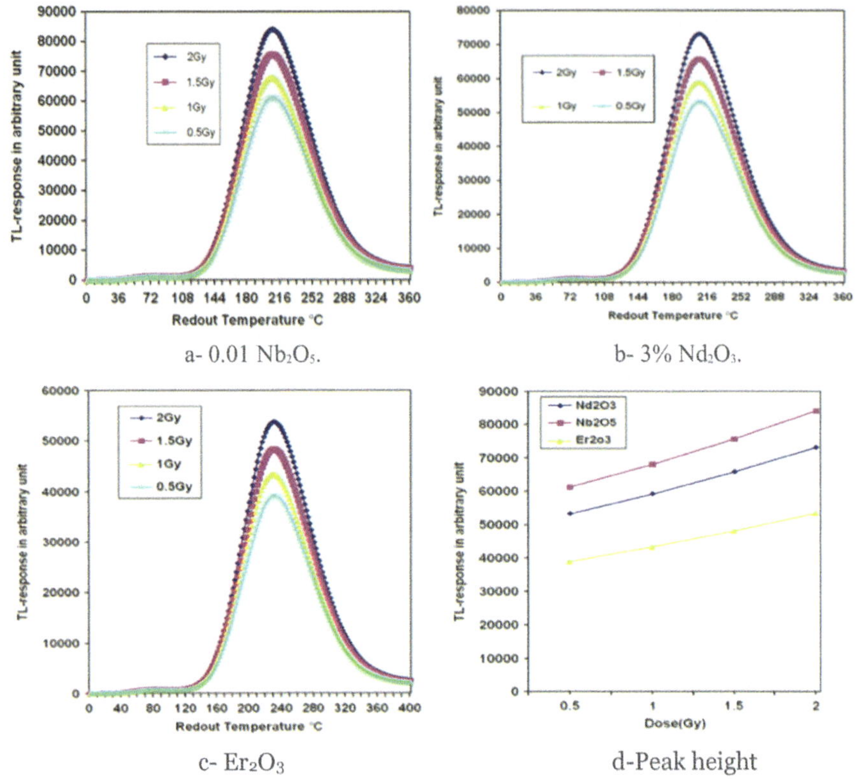


**Fig. 4.2** **a–d** TL response vs. dose of tellurite glass doped with $Nd_2O_3$, $Nb_2O_5$, and $Er_2O_3$ for irradiation with $^{60}Co$ photons [3] with permission. **a** 0.01 $Nb_2O_5$. **b** 3% $Nd_2O_3$. **c** $Er_2O_3$. **d** Peak height

micro-Raman spectrometer [8] as shown in Fig. 4.4a–e. The obtained luminescence intensity ratio was correlated with:

1. The concentration of rare-earth ions,
2. Short distance $r$ (Ln–Ln),
3. Glass density.

This behavior can be interpreted by calculating the number of bonds ($n_b$) per unit volume,

$$n_b = \frac{N_a}{V_M} \sum_i x_i\, n_i \tag{4.3}$$

where $n_i$ is the number of bridging bonds per cation in the $i$th oxide component, and $N_a$ is Avogadro's number. The calculated values of ($n_b$) for the undoped sample are $103.382 \times 10^{21}$ cm$^{-3}$, and its value for the glass doped with $Nd_2O_3$, $Sm_2O_3$,

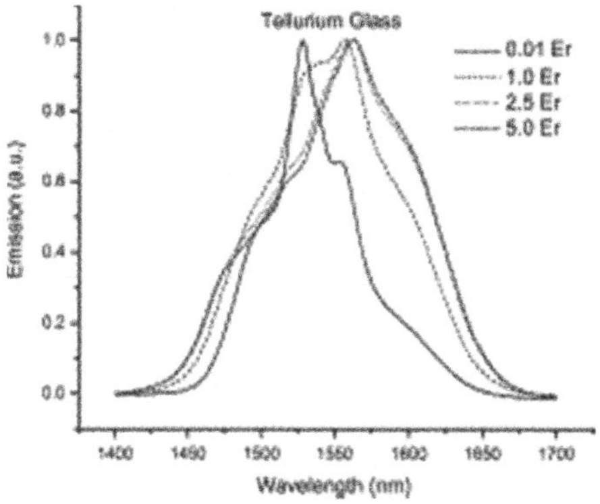


a- Spectra of $TeO_2$–$TiO_2$– $WO_3$ glass for different concentrations of $Er^{3+}$ ions [4] with permission.

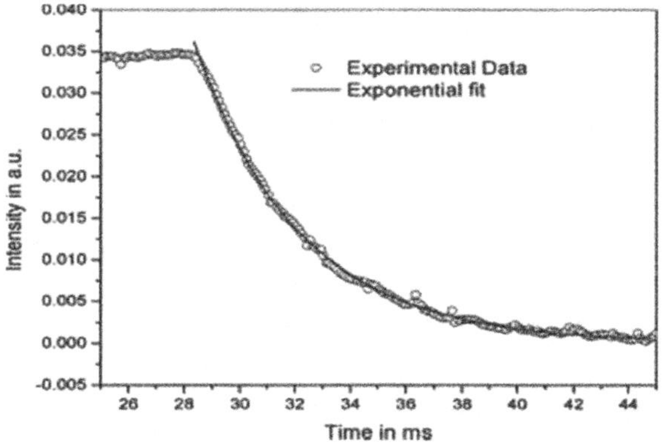


b- Decay curve of $^4I_{13/2}$ level of $Er^{3+}$ in $TeO_2$–$TiO_2$– $WO_3$ glass [4] with permission.

**Fig. 4.3 a–d** Up converted fluorescence emission spectra of different concentrations (mol%) of $Er^{3+}$ in $TeO_2$–$TiO_2$–$WO_3$ glasses [4] with permission. **a** Spectra of $TeO_2$–$TiO_2$–$WO_3$ glass for different concentrations of $Er^{3+}$ ions [4] with permission. **b** Decay curve of $^4I_{13/2}$ level of $Er^{3+}$ in $TeO_2$–$TiO_2$–$WO_3$ glass [4] with permission. **c** Spectral width and decay time of $^4I_{13/2} \rightarrow {}^4I_{15/2}$ transition with concentration of $Er^{3+}$ ions [4] with permission. **d** The energy level diagram for $Er^{3+}$ ions, infrared excitation of 975 nm

and $Er_2O_3$ raised to $109.944 \times 10^{21}$ and $110.209 \times 10^{21}$ and $110.057 \times 10^{21}$ $cm^{-3}$, respectively [8].

The doped rare-earth ions produced the light peaks that were seen. These luminescence bands are the result of transitions that happened after excited state absorption. The potential for these tellurite glasses doped with $Nd_2O_3$, $Sm_2O_3$, and $Er_2O_3$ in the realm of laser applications has been raised [8].

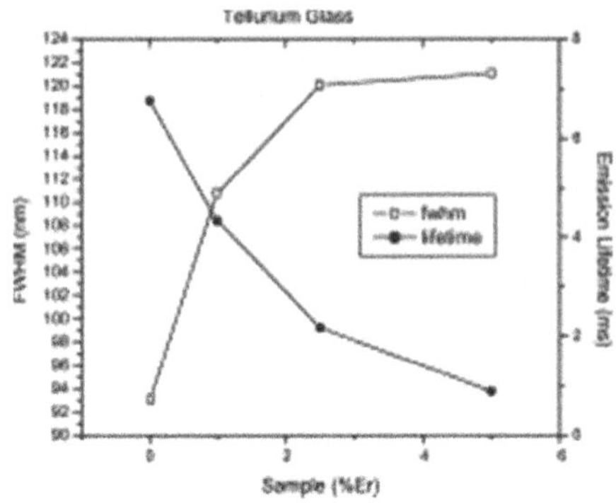


c-Spectral width and decay time of $^4I_{13/2} \rightarrow {}^4I_{15/2}$ transition with concentration of $Er^{3+}$ ions [4] with permission.

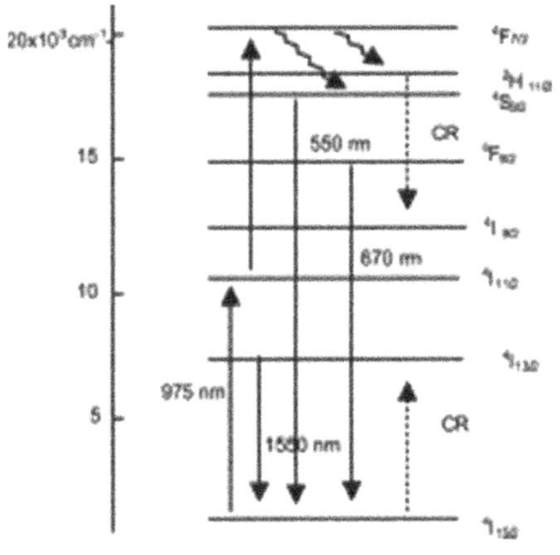


d-The energy level diagram for $Er^{3+}$ ions, infrared excitation of 975 nm.

**Fig. 4.3** (continued)

## 4.3 Emission and Absorption

Lithium-borotellurite glasses that were doped with $Sm^{3+}$ (1.0%) and $Dy^{3+}$ (1.0%) were analyzed spectroscopically [7]. $Dy^{3+}$ glass has been used to obtain the blue (4F9/26H 15/2) and yellow (4F9/26H 13/2) emission bands. The lifetimes of the emissions of the $Sm^{3+}$ glass (4G5/2 → 6H 5/27/29/2 and 11/2) at an excitation of 401 nm were in the range of 0.47–0.81 ms. The stimulated emission of $\sigma_p^E$ cross sections of the detected emission transitions has also been calculated, and the values ranged from 0.38 to 1.20 1020 $cm^2$ for $Sm^{3+}$ and from 0.66 to 1.39 1020 $cm^2$ for $Dy^{3+}$-doped lithium-borotellurite glass. They concluded that the $Sm^{3+}$-doped LiBTe glasses emit a vivid orange (4G 5/2 6H 7/2), a red (4G 5/2 6H 9/2), and a yellow (4G 5/2 6H 5/2) emission. The $Dy^{3+}$ glass has been found to emit both blue (4F9/26H

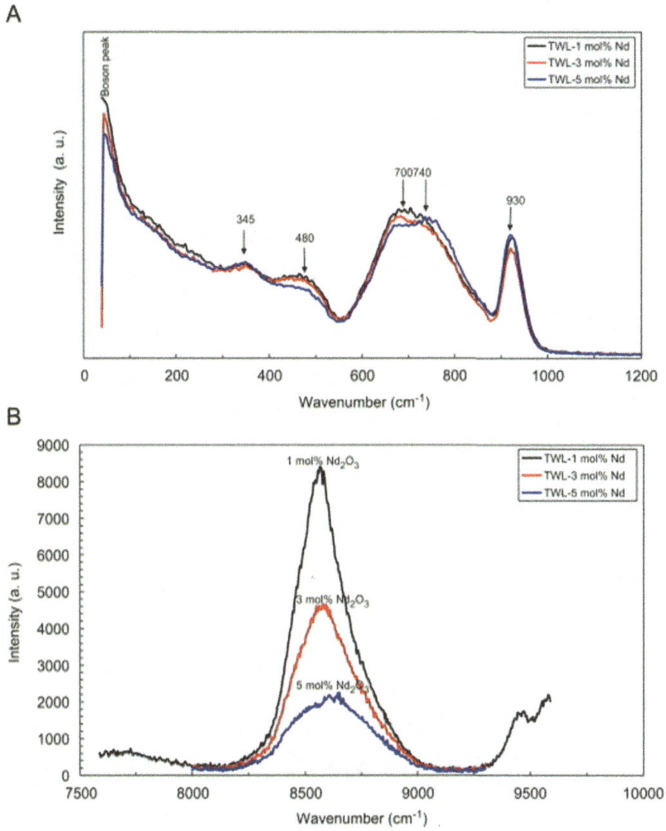


Signals of (A) Raman and (B) luminescence of $Nd_2O_3$ doped $TeO_2$–$WO_3$–$Li_2O$ glass under 476.5 nm excitation of an argon laser [8] with permission.

(a)

**Fig. 4.4** **a–e** Variation in optical polarizability with a short distance between two identical rare-earth ions [8]. With permission. **a** Signals of (A) Raman and (B) luminescence of $Nd_2O_3$-doped $TeO_2$–$WO_3$–$Li_2O$ glass under 476.5 nm excitation of an argon laser [8] with permission. **b** Signals of (A) Raman and (B) luminescence of $Sm_2O_3$-doped $TeO_2$–$WO_3$–$Li_2O$ glass under 514.5 nm excitation of an argon laser. **c** Signals of (A) Raman and (B) luminescence of $Er_2O_3$-doped $TeO_2$–$WO_3$–$Li_2O$ glass under 476.5 nm excitation of an argon laser. **d** Excitation power at 514.5 nm—luminescence integrated intensity of emission at 1670, 2740, and 3948 $cm^{-1}$ in a sample containing 1 mol% $Sm_2O_3$

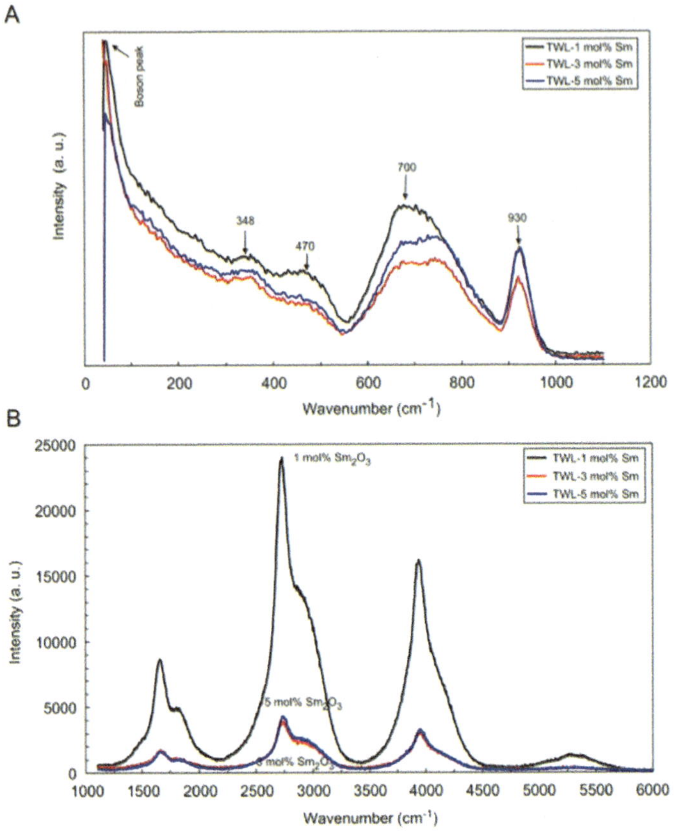


Signals of (A) Raman and (B) luminescence of $Sm_2O_3$ doped $TeO_2$–$WO_3$–$Li_2O$ glass under 514.5 nm excitation of an argon laser.

(b)

**Fig. 4.4** (continued)

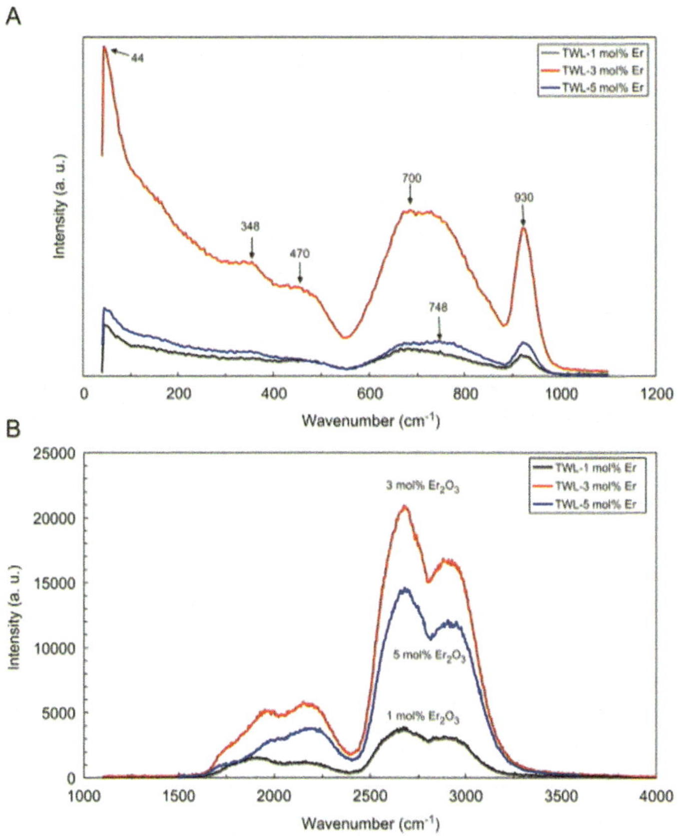


Signals of (A) Raman and (B) luminescence of $Er_2O_3$ doped $TeO_2$–$WO_3$–$Li_2O$ glass under 476.5 nm excitation of an argon laser.

(c)

**Fig. 4.4** (continued)

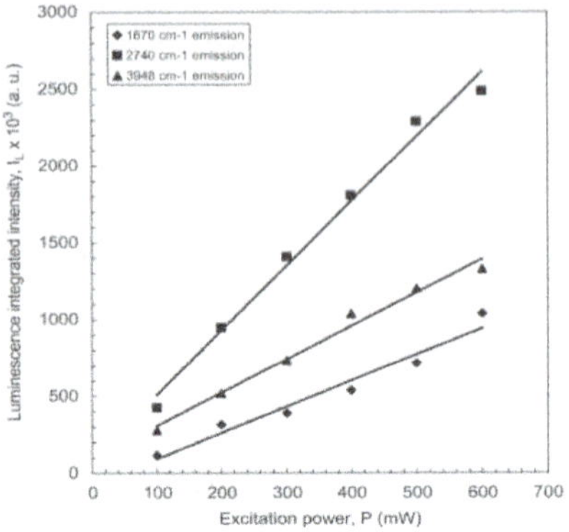


Excitation power at 514.5 nm - luminescence integrated intensity of emission at 1670, 2740, and 3948 cm$^{-1}$ in a sample containing 1 mol% $Sm_2O_3$.

(d)

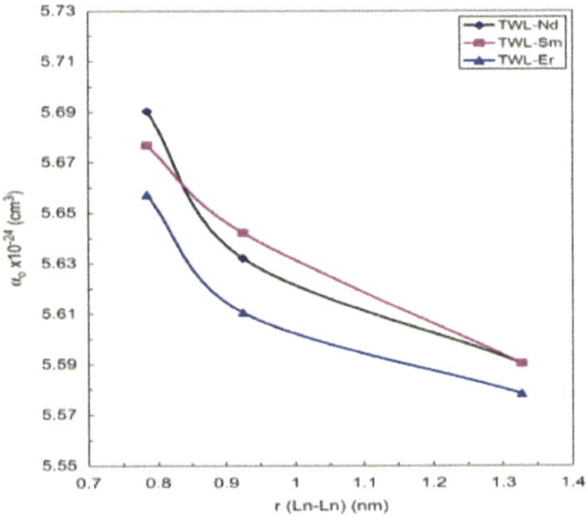


(e)

**Fig. 4.4** (continued)

15/2) and yellow (4F9/26H 13/2) emissions, and both of these emissions are equally intense in this glass.

Such glasses could be referred to as phosphorescent optical materials since the emission transitions' lifetimes are measured in milliseconds. The detected orange emission transition for $Sm^{3+}$ has a stimulated emission cross section ($\sigma_p^E$) with a minimum cross-section value of 0.38 10–20 cm$^2$, while the blue emission band for $Dy^{3+}$ has a cross-section value of 0.66 10–20 cm$^2$.

Based on their spectrum properties, both of these optical glasses are therefore very interesting as new optical materials for use in illuminating systems.

## 4.4 White Light Source and Optical Thermometry

White light-emitting diode research has gained popularity in recent years [9–37]. Researchers looked at the possibility of $Tm^{3+}/Er^{3+}/Sm^{3+}$ tri-doped (TLZ) glasses for temperature sensors that produce colored emissions as well as for producing white light sources [38]. Glass compositions were in the form:

TLZ–0.2 Sm: {$65TeO_2$–$15Li_2O$–20ZnO}–$Er_2O_3$ (0.5 mol%)–$Tm_2O_3$ (0.2 mol%)–$Sm_2O_3$ (0.2 mol%),
TLZ–0.5Sm: „ „ „–$Sm_2O_3$ (0.5 mol%),
TLZ–1.0Sm: „ „ „–$Sm_2O_3$ (1.0 mol%).

The photoluminescence (PL) excitation spectra Fig. 4.5a–d of the TLZ0.5Sm sample associated with the emissions at 459 nm (blue line), 530 nm (green line), and 604 nm (red line) are depicted in:

(a) Excitation spectra of the TLZ0.5Sm sample in the spectra range from 340 to 410 nm corresponding to three emission wavelengths: 459 nm of the $Tm^{3+}$ ions (blue curve), 530 nm of the $Er^{3+}$ ions (green curve) and 604 nm of the $Sm^{3+}$ ions (red curve),
(b) Emission spectra of the TLZ0.2Sm, TLZ5.0Sm, and TLZ1.0Sm samples, under excitation at 360 nm. The emission spectra were obtained using a KV399 filter,
(c) An energy level diagram of a tri-doped $Tm^{3+}/Er^{3+}/Sm^{3+}$ sample showing potential energy transfer paths.

## 4.5 Impact of Rare-Earth Nanoparticles on Up-Conversion, Down-Conversion, and Thermal Properties of Tellurite Glasses

The cohesion of atoms and the volume of the unit cell are two factors that have already been mentioned (Chap. 1) as controlling the size-dependent density of nanoparticles [39–41]. The current section will be concentrated on the up-conversion, optical efficiency, and thermal characteristics of tellurite glasses doped with rare-earth nanoparticles [42]. The tellurite glass series' up-conversion emission spectra were excited at an 800 nm wavelength [42]. There are various factors responsible for the shift in emission peaks [43–45].

The melting temperature, absorbance, and nonlinearity of silicate glass are all high [46]. Tellurite glasses have a wide range of optical applications, including yellowish-green optical devices, laser applications, visible upconverters, and tuneable display devices, according to earlier studies [47–50]. Tellurite glass appears to be one of the unique and high-demand materials for its interesting physical properties [51–56].

Moreover, tellurite glass has undergone substantial development for use in a range of optical applications, such as glass sensors, light-emitting diodes (LEDs), fiber optics, and optoelectronics [57]. The $Er^{3+}$, a doped fiber amplifier (EDFA) device,

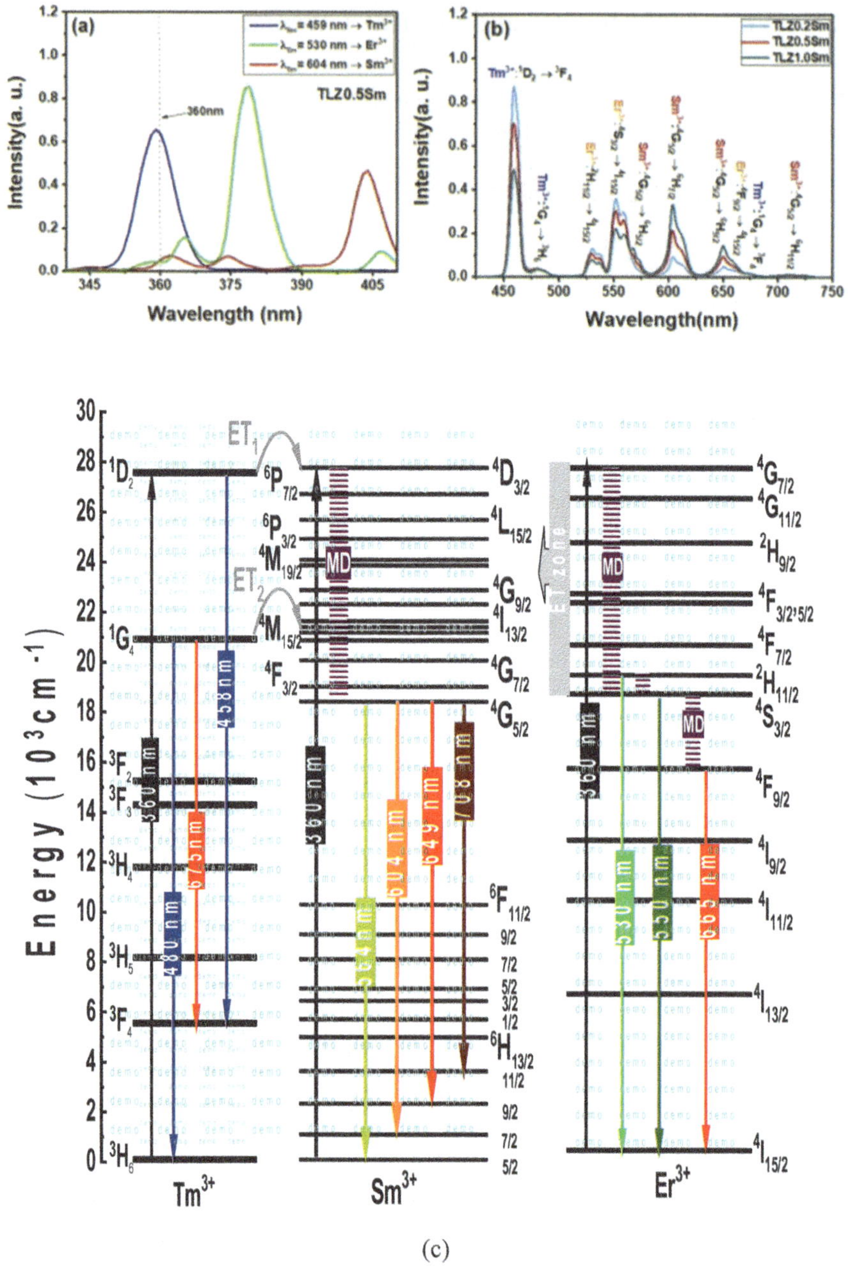

**Fig. 4.5** White light source and optical thermometry of $Tm^{3+}/Er^{3+}/Sm^{3+}$ tri-doped $TeO_2$–$Li_2O$–ZnO (TLZ) [38] with permission

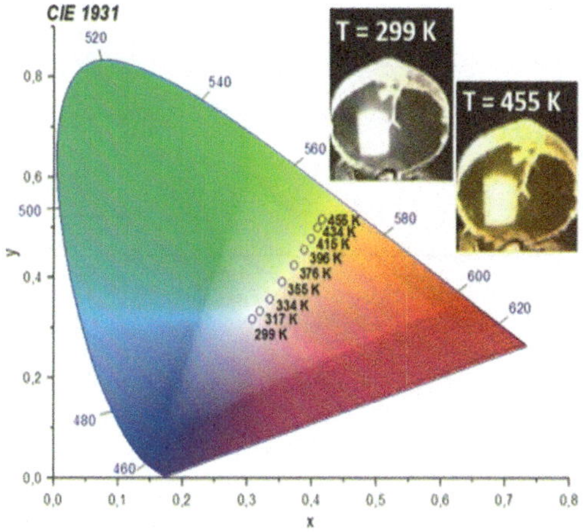


| T(K) | x | y |
|---|---|---|
| 299 | 0.308 | 0.317 |
| 317 | 0.319 | 0.333 |
| 334 | 0.335 | 0.356 |
| 355 | 0.354 | 0.390 |
| 376 | 0.372 | 0.423 |
| 396 | 0.388 | 0.454 |
| 415 | 0.399 | 0.476 |
| 434 | 0.410 | 0.498 |
| 455 | 0.416 | 0.514 |

(d)

**Fig. 4.5** (continued)

is one of the applications for rare-earth oxides, which are utilized to improve optical communication. The use of erbium oxide to improve the optical qualities of tellurite glass has been extensively studied [58]. A study found that laser glass's optical performance is significantly influenced by particle size [59].

The up-conversion capabilities of tellurite glasses doped with erbium nanoparticles for highly effective laser glass have been measured in [60]. The up-conversion emission band under stimulation of 800 nm is shown in the inset of Fig. 4.6a, b. The divergence of the 4f configuration was the primary cause of the observed emission peak, according to the Russell–Saunders coupling [61–66]. Therefore, the excitation of $Er^{3+}$ ions can be used to observe the emission at the ultraviolet and infrared areas. Tellurite glass doped with erbium NPs has a strong emission peak at 0.05 mol fraction, as seen in Fig. 4.6a, b. The significant number of excited $Er^{3+}$ nanoparticle ions at 0.05 mol fraction is what causes the strong emission peak.

The presence of an impurity band produced by the doping concentrations is what causes the red shift in Fig. 4.6b. Another explanation for the redshift is that it results from possible oscillations that cause vacancies, impurities, dangling bonds, and structural flaws to emerge [66]. The J-O theory calculation can be obtained in earlier studies [67, 68]. In Fig. 4.6a–c, the relationship between the trend of spectral parameters and the concentration of erbium NPs is shown. Due to their exceptional optical performance, the current glasses provide appropriate materials for laser applications.

In 2020, emphasis on the up-conversion emission of the laser glass excited at 800 nm was found in the ultraviolet region for neodymium NPs-doped $\{[(TeO_2)0.7(B_2O_3)0.3]0.7\ (ZnO)0.3\}_{1-x}\ (Nd_2O_3\ NPs)_x$ ($x = 0.005$–$0.05$ mol%) [69] as shown in Fig. 4.7a, b. According to an analysis in Ref. [70–74], the emission

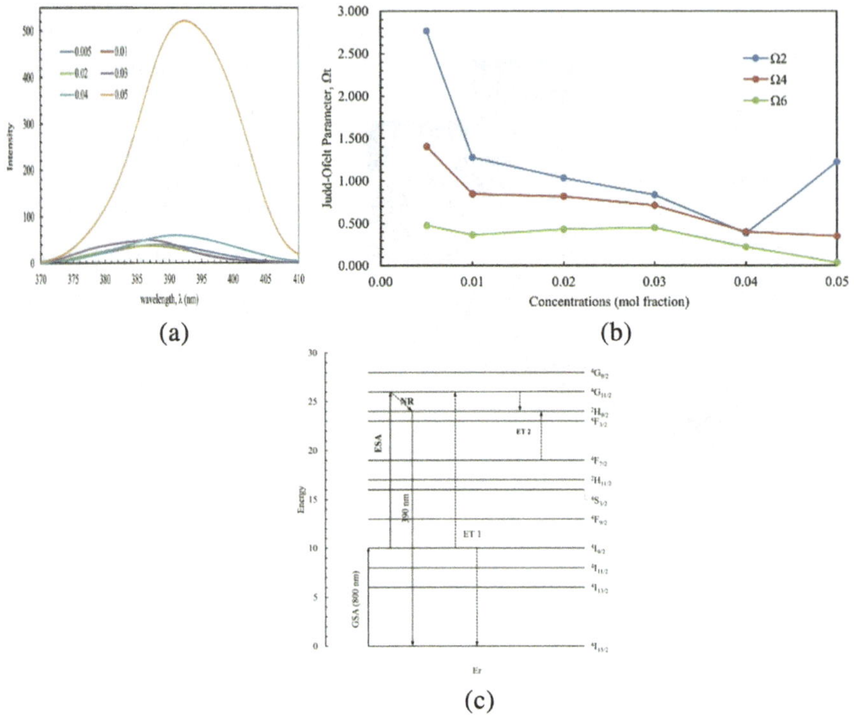


**Fig. 4.6** **a–c** Up-conversion emission spectra and Judd–Ofelt (J–O) properties of erbium nanoparticles in $\{(TeO_2)\ (B_2O_3)\ (ZnO)_{0.3}\}_y$ ($Er_2O_3$NPs), glass [60]. With permission

peaks of neodymium NPs exhibit blue shifts at 0.01 and 0.03 mol fractions and a red shift at 0.02 mol fraction. Moreover, both thermal and structural properties of the prepared glass series $Er_2O_3$ NPs and $Ag_2O$-doped bio-silica borotellurite glass system after to extraction of silicate from the rice husk have been shown that: the main thermal parameters $T_g$, $T_c$, $T_s$, and Hruby parameter increased from 469 to 495, 700 to 768, 231 to 273 °C, and 0.469 to 0.644 [75].

## 4.6 Tellurite Glasses, Tellurite Glass Ceramics, and Tellurite Ceramics for Different Purposes

The mechanism of down-conversion in $Er^{3+}/Yb^{3+}$-doped fluorotellurite glasses to increase the efficiency of c-Si PV cells has been reported in 2022 as shown in Fig. 4.8 [76]. The authors concluded that these glasses could increase the spectrum sensitivity of crystalline silicon photovoltaic cells by converting visible photons into NIR emissions at 1000 nm.

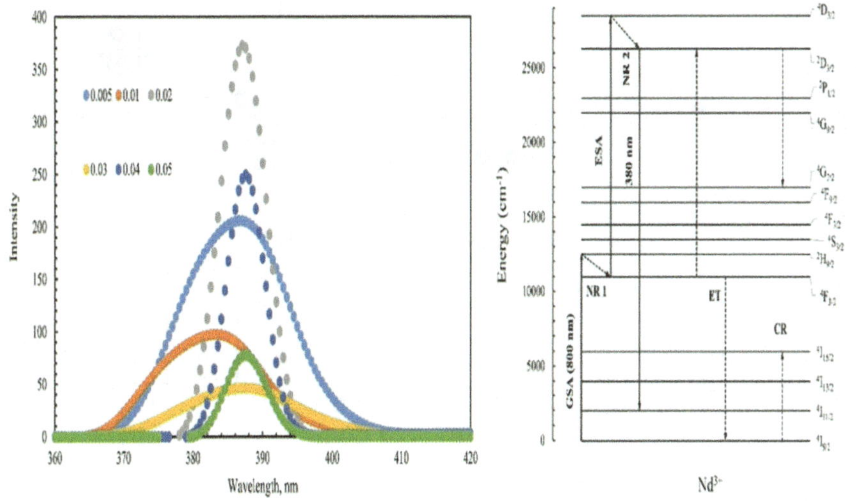


(a) (b)

**Fig. 4.7** **a, b** Up-conversion spectra and schematic diagram of ($Nd_2O_3$–NPs)-doped tellurite glasses ($TeO_2$)–($B_2O_3$) (ZnO)–($Nd_2O_3$–NPs)$_x$ [69]. With permission

Particularly, Sailaja et al. [77] studied the near-IR photoluminescence studies of $Nd^{3+}$ ions-doped SrO–$Al_2O_3$–$BaCl_2$–$B_2O_3$–$TeO_2$ glasses for fiber amplifier applications. The energy level diagram of $Nd^{3+}$ ion-doped OCBT glasses has been illustrated in Fig. 4.9a–d [77]. The OCBTNd with 1 mol% of $Nd^{3+}$ ions-doped OCBT glass sample gave the highest quantum efficiency. It may be inferred from the fruitful results that OCBTNd-1.0 glass finds lots of use in solid-state laser technology.

Furthermore, Morova et al. [78] studied the tunable continuous-wave laser operation of tellurite glass doped with $Tm^{3+}$ ion near 2 μm as shown in Fig. 4.10. The authors stated that the emission cross section was $6.8 \times 10^{-21}$ $cm^2$ 1 at 1926 nm.

More than that effects of melting temperature on color-changing in germano-tellurite niobate glass have been reported by Wang et al. [79] as shown in Fig. 4.11a–e [79]. This work will provide a new perspective on clarifying the effects of melting temperature on color-changing in the germano-tellurite niobate glass. Recently, Tang et al. [80] studied the 2–3 μm mid-IR luminescence of $Ho^{3+}/Yb^{3+}$ co-doped chloride-modified fluorotellurite glass as shown in Fig. 4.12. Glasses in the form of $(100 - x)$ ($AlF_3$–$MgF_2$–$CaF_2$–$SrF_2$–$YF_3$–$TeO_2$)–$xBaX_2$ ($X$ = F, Cl) were prepared by the conventional melting technique. The achieved chloride in fluorotellurite glass reveals excellent performance in mid-infrared luminescence. So, this glass got potential applications in mid-IR fiber amplifiers and laser devices. After that, Li et al. [81] studied the broadband NIR radiative transitions in $Er^{3+}/Tm^{3+}$ co-doping tellurite glass material as represented in Fig. 4.13. It has been concluded that these glasses are a promising broadband gain medium, and embedding Ag NPs is an effective scheme to improve the radiative characteristics.

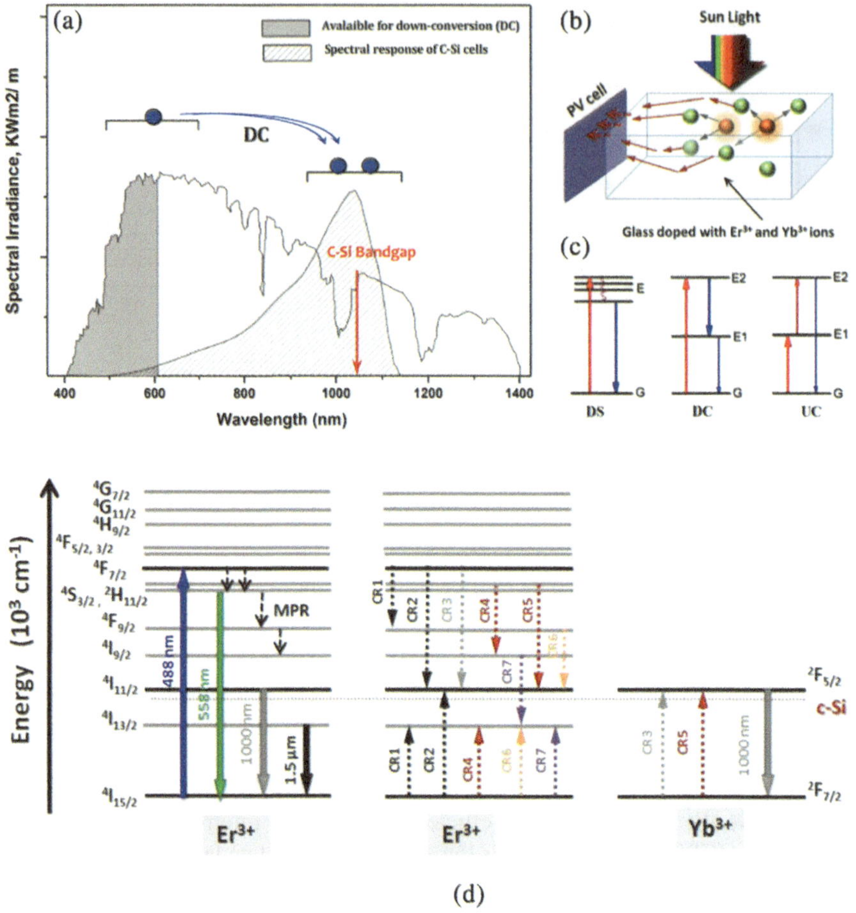

**Fig. 4.8** **a** AM1.5 solar spectrum and spectral response of c-Si cells, **b** LSC based on glass co-doped with erbium and ytterbium ions, **c** Down-shifting (DS), down-conversion (DC), and up-conversion (UC) processes, **d** Energy level diagrams of $Yb^{3+}$ and $Er^{3+}$ ions, the processes implicate in down-conversion (DC) emissions of TNZF glass systems under the excitation at 488 nm [76] With permission

Transparent tellurite glass ceramics for photonics applications were described by Patra et al. [82]. One of the nontraditional oxide glass media that has received the most research is the tellurite glass system. With unique features as illustrated in Fig. 4.14, transparent tellurite glass ceramics made by precipitating non-centro symmetric nano-crystallites dispersed in a glass matrix have surely expanded the scope of their photonic applications. The precipitated crystallite volume fraction, size, and shape of a multicomponent transparent glass ceramic material may be accurately optimized, the concentration of the nucleating agent can occasionally be adjusted, and the heat treatment process is closely monitored. Thus, all distinctive

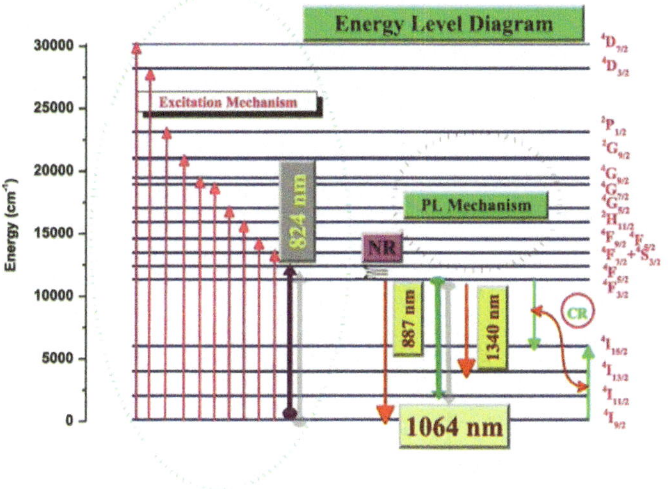


**Fig. 4.9** Energy level diagram of $Nd^{3+}$ ion-doped OCBT glasses [77] With permission

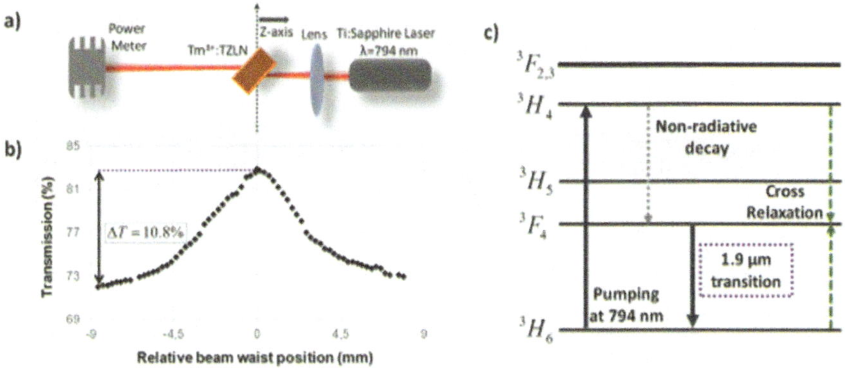


**Fig. 4.10** **a** Experimental setup for the z-scan absorption saturation measurement at 794 nm, **b** Measured variation of the transmission for the $Tm^{3+}$: TZLN glass as a function of the relative sample position at the incident pump power of 665 mW, **c** Energy level diagram of the $Tm^{3+}$ ion showing the optical transitions for absorption at 794 nm and lasing near 1.9 μm [78]. With permission

functionalities of glass ceramics were controlled by the characteristics of precipitated crystalline phases and the final composition of the residual glassy phase.

Therefore, it is crucial to conduct a thorough analysis of the micro and nanostructure as well as the morphology of the crystalline phase. This has mostly been done in the present review, which shows that even a small distortion in the crystallite phase's center of symmetry increases optical nonlinearity. Additionally, the development of an asymmetric environment surrounding RE ions alters the emission peak form

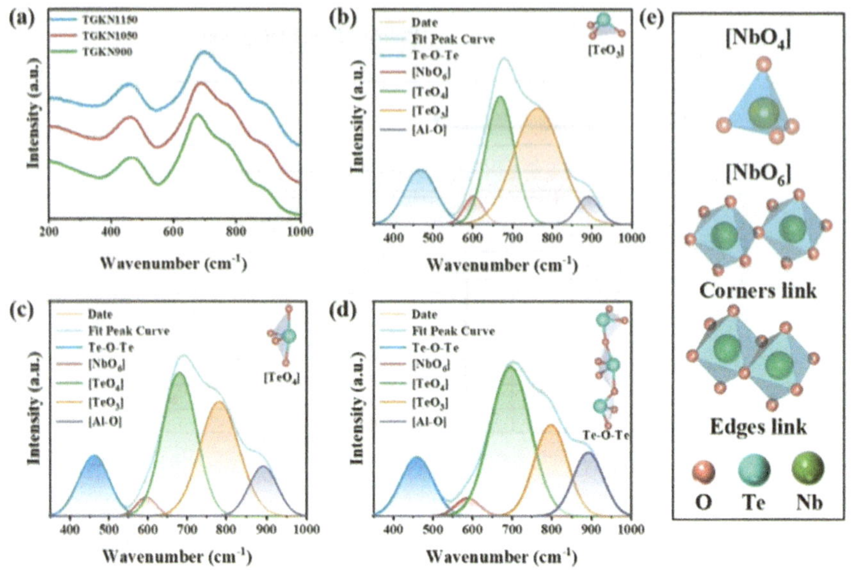


**Fig. 4.11** **a–e** Raman spectra of TGKN of glass series (**a**); Gaussian peak deconvolution of the Raman spectra obtained for TGKN900 (**b**), TGKN1050 (**c**), TGKN1150 (**d**), **e** schematic illustration for the possible connection among $Nb_2O_5$ [79]. With permission

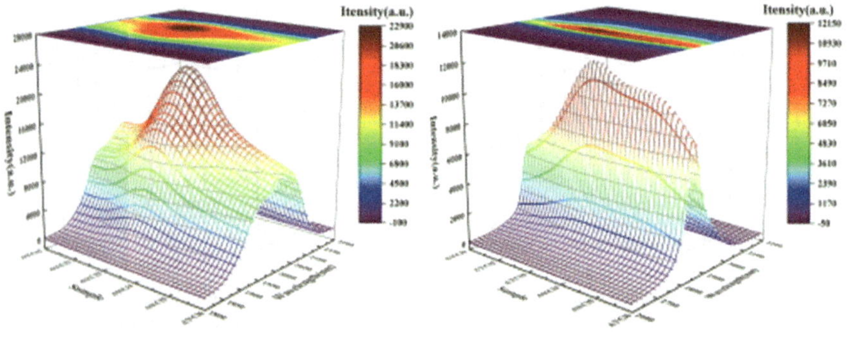


**Fig. 4.12** 2–3 μm mid-infrared luminescence of $Ho^{3+}/Yb^{3+}$ co-doped chloride-modified fluorotellurite glass [80]. With permission

(sharp, broad, or splitting) and intensity, opening up new possibilities for the material's potential applications. They have given a thorough explanation of the various procedures that must be taken throughout the heat treatment process to control the ceramicization of glass, allowing for the fabrication of far more creative transparent glass ceramics for use in the future.

An in-depth discussion is given on recently discovered crystallization methods such as congruent crystallization and reverse classical nucleation theory (CNT) heat

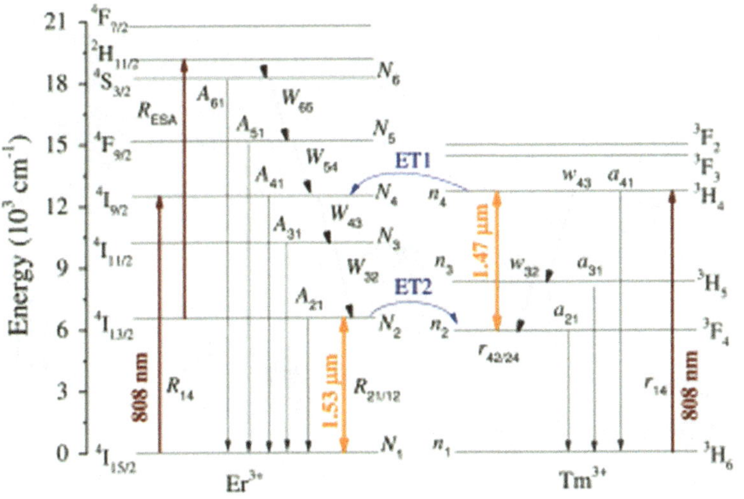


**Fig. 4.13** Schematic energy level transition diagram and energy transfer of $Er^{3+}$ and $Tm^{3+}$ co-doping tellurite glass under the excitation of 808 nm LD [81]. With Permission

treatment, which is used to create fiber amplifiers and lasers. The future aspect section of this article intends to provide an outlook on the tellurite-based transparent glass ceramics systems for different applications. Along with the use of QDs at tellurite glass composite materials like WLED, random laser, etc., which is typically not possible from other matrices, this review also emphasizes the suitability of RE ion-activated transparent tellurite glass ceramics for advanced photonic applications such as bulk laser, fiber laser, fiber amplifier, and up-conversion laser material. Transparent tellurite glass ceramics have a lot of potential, yet there are presently just a few real-world uses for them. This review has also emphasized the rising application areas for transparent tellurite glass ceramics to strengthen all the impending application domains, which are all amateur. Figure 4.14 shows some applications and challenges for transparent tellurite glasses- ceramics [82].

Dolhen et al. [83] achieved the $Nd^{3+}$—doped transparent tellurite ceramics bulk lasers as represented in Fig. 4.15. Therefore, this work intends to evaluate the laser characteristics of such transparent tellurite ceramics made by fully congruent crystallization from the parent glass since there are no studies dealing with this. The developed bulk tellurite ceramics are characterized in the first section of the study, which also presents and discusses structural and microstructural data as well as optical transparency and photoluminescence properties. The lasing emission in both the continuous wave (cw) and pulsed regimes are reported in the second section. Details concerning the laser beam's spatial form, its emission spectral properties, the impact of the output couplers used, and the effect on the measured laser efficiency are shown in Fig. 4.15.

It has been concluded that $Nd^{3+}$ doped at 1 mol% using a congruent crystallization procedure, translucent ceramics made of $75TeO_2$–$12.5Bi_2O_3$–$12.5Nb_2O_5$ glasses

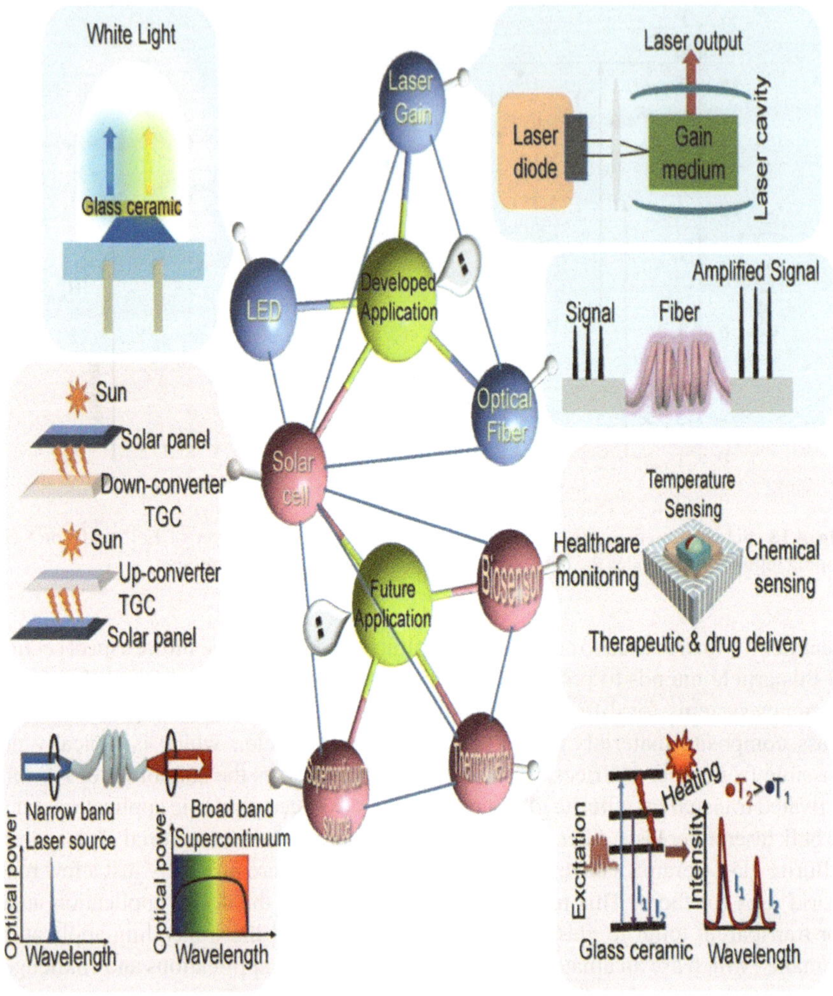


**Fig. 4.14** Schematic for developed and future application of transparent tellurite glass ceramics [82] With permission

were successfully produced at 510 °C in an air environment. It was established that careful temperature control was required to prevent the cubic $Bi_{0.8}$ $Nb_{0.8}$ $Te_{2.4}O_8$ Ia-3 crystal phase from decomposing into the orthorhombic Bi–Nb–$Te_2O_8$ crystal phase at the surface of the heating process. In both the (cw) and gain-switching regimes, laser emission was seen. In the (cw) regime, a maximum output power of around 28.5 mW was achieved for a laser threshold of about 67 mW and efficiency and slope efficiency of about 22.5 or 50%, respectively. In contrast, pulsed circumstances demonstrated a peak power of more than 6.5 W at a repetition rate of 728 kHz.

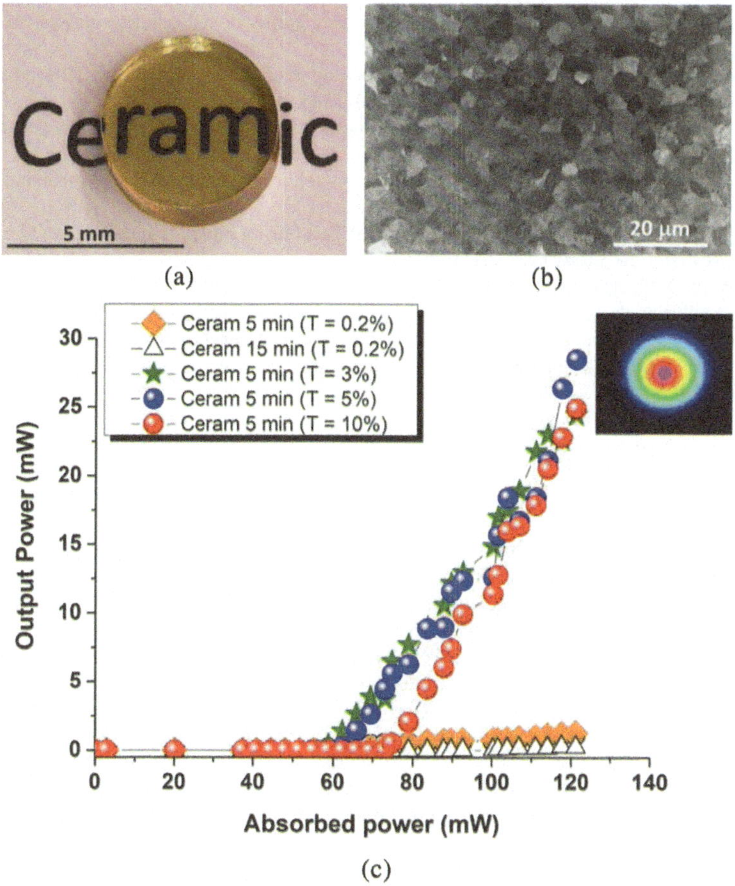


**Fig. 4.15** **a** Bulk tellurite ceramic, **b** SEM observation of the polished surface, **c** Laser output power at 1064 nm versus the input pump power at 808 nm for different output couplers and two ceramic samples synthesized after 5 min and 15 min at 510 °C [83]

It is interesting to re-answer the two questions in the preface according to the importance of these glasses, glasses and glass ceramics as a condensed matter physics with unique physical properties. Also, uuniversal power-law of terahertz optical properties, near-infrared luminescence property, the effect of $Tm_2O_3$ addition on the physical, structural, elastic, and improving emission performance.

## References

1. Abdel-Kader, A., et al. 1994. Thermoluminescence dosimetry of rare-earth doped tellurite phosphate glasses. *Materials Chemistry and Physics* 36 (3–4): 365–370.

2. Elkholy, M. 2003. Thermoluminescence for rare-earths doped tellurite glasses. *Materials Chemistry and Physics* 77 (2): 321–330.
3. El-Mallawany, R., and H. Diab. 2012. Improving dosimetric properties of tellurite glasses. *Physica B: Condensed Matter* 407 (17): 3580–3585.
4. El-Mallawany, R., et al. 2004. Study of luminescence properties of $Er^{3+}$-ions in new tellurite glasses. *Optical Materials* 26 (3): 267–270.
5. Shen, S., M. Naftaly, and A. Jha. 2002. Tungsten–tellurite—a host glass for broadband EDFA. *Optics Communications* 205 (1–3): 101–105.
6. Himei, Y., et al. 1994. Coordination change of Te atoms in binary tellurite glasses. *Journal of Non-Crystalline Solids* 177: 164–169.
7. Hussain, N.S., et al. 2009. Absorption and emission analysis of $RE^{3+}$ ($Sm^{3+}$ and $Dy^{3+}$): Lithium boro tellurite glasses. *Journal of Nanoscience and Nanotechnology* 9 (6): 3672–3677.
8. Hager, I., R. El-Mallawany, and A. Bulou. 2011. Luminescence spectra and optical properties of $TeO_2$–$WO_3$–$Li_2O$ glasses doped with Nd, Sm and Er rare earth ions. *Physica B: Condensed Matter* 406 (4): 972–980.
9. Gökçe, M., and D. Koçyiğit. 2019. Spectroscopic investigations of $Dy^{3+}$ doped borogermanate glasses for laser and wLED applications. *Optical Materials* 89: 568–575.
10. Ji, C., et al. 2021. $Sm^{3+}$/$Pr^{3+}$ biactivated $Ca_3Y_2Ge_3O_{12}$: 0.04 $Sm^{3+}$: $Pr^{3+}$ red phosphor with high thermal stability for low correlated temperature WLED. *Journal of Luminescence* 232: 117775.
11. Chen, Q., et al. 2021. Color-tunable $Eu^{3+}$-or $Sm^{3+}$-doped perovskite phosphors as optical temperature-sensing materials. *Optical Materials* 111: 110585.
12. Pandey, N., et al. 2019. Efficient fluorescence quenching of 5-aminoquinoline: Silver ion recognition study. *Journal of Luminescence* 205: 475–481.
13. Gonçalves, A., et al. 2018. Luminescence and upconversion processes in $Er^{3+}$-doped tellurite glasses. *Journal of Luminescence* 201: 110–114.
14. Chen, W., et al. 2018. Sr2GdF7: $Tm^{3+}$/$Yb^{3+}$ glass ceramic: A highly sensitive optical thermometer based on FIR technique. *Journal of Alloys and Compounds* 735: 2544–2550.
15. Selvaraju, K., and K. Marimuthu. 2013. Structural and spectroscopic studies on concentration dependent $Sm^{3+}$ doped boro-tellurite glasses. *Journal of Alloys and Compounds* 553: 273–281.
16. El-Mallawany, R. 1992. The optical properties of tellurite glasses. *Journal of Applied Physics* 72 (5): 1774–1777.
17. Sayyed, M., and R. El-Mallawany. 2017. Shielding properties of (100–x) $TeO_2$–(x) $MoO_3$ glasses. *Materials Chemistry and Physics* 201: 50–56.
18. Elazoumi, S., et al. 2018. Effect of PbO on optical properties of tellurite glass. *Results in Physics* 8: 16–25.
19. El-Mallawany, R., et al. 1994. Elastic constants of semiconducting tellurite glasses. *Materials Chemistry and Physics* 37 (3): 295–298.
20. Walas, M., et al. 2018. Tailored white light emission in $Eu^{3+}$/$Dy^{3+}$ doped tellurite glass phosphors containing $Al^{3+}$ ions. *Optical Materials* 79: 289–295.
21. Uma, V., et al. 2018. Luminescence and energy transfer studies on $Sm^{3+}$/$Tb^{3+}$ codoped telluroborate glasses for WLED applications. *Journal of Molecular Structure* 1151: 266–276.
22. Denker, B., et al. 2018. Rare-earth ions doped zinc-tellurite glass for 2 ÷ 3 μm lasers. *Applied Physics B* 124: 1–8.
23. Yuan, J., and P. Xiao. 2018. Compositional effects of $Na_2O$, $GeO_2$, and $Bi_2O_3$ on 1.8 μm spectroscopic properties of $Tm^{3+}$ doped zinc tellurite glasses. *Journal of Luminescence* 196: 281–284.
24. Bednarkiewicz, A., et al. 2020. Standardizing luminescence nanothermometry for biomedical applications. *Nanoscale* 12 (27): 14405–14421.
25. Jaque, D., and F. Vetrone. 2012. Luminescence nanothermometry. *Nanoscale* 4 (15): 4301–4326.
26. Yang, P., et al. 2019. Realizing emission color tuning, ratiometric optical thermometry and temperature-induced redshift investigation in novel $Eu^{3+}$-doped $Ba_3La$ $(VO_4)_3$ phosphors. *Dalton Transactions* 48 (29): 10824–10833.

27. Ximendes, E.C., et al. 2017. In vivo subcutaneous thermal video recording by supersensitive infrared nanothermometers. *Advanced Functional Materials* 27 (38): 1702249.
28. Gharouel, S., et al. 2018. Fluorescence intensity ratio and lifetime thermometry of praseodymium phosphates for temperature sensing. *Journal of Luminescence* 201: 372–383.
29. Serkova, N.J. 2017. Nanoparticle-based magnetic resonance imaging on tumor-associated macrophages and inflammation. *Frontiers in Immunology* 8: 590.
30. Liu, H., et al. 2015. Intracellular temperature sensing: An ultra-bright luminescent nanothermometer with non-sensitivity to pH and ionic strength. *Scientific Reports* 5 (1): 14879.
31. Gao, M., et al. 2017. Tumor acidity-activatable TAT targeted nanomedicine for enlarged fluorescence/magnetic resonance imaging-guided photodynamic therapy. *Biomaterials* 133: 165–175.
32. Kramer, N.J., K.S. Schramke, and U.R. Kortshagen. 2015. Plasmonic properties of silicon nanocrystals doped with boron and phosphorus. *Nano Letters* 15 (8): 5597–5603.
33. Stark, A.M., and S. Way. 1974. The use of thermovision in the detection of early breast cancer. *Cancer* 33 (6): 1664–1670.
34. Ortgies, D.H., et al. 2019. Infrared fluorescence imaging of infarcted hearts with $Ag_2S$ nanodots. *Nano Research* 12: 749–757.
35. Santos, H.D., et al. 2018. In vivo early tumor detection and diagnosis by infrared luminescence transient nanothermometry. *Advanced Functional Materials* 28 (43): 1803924.
36. Ximendes, E.C., et al. 2017. In vivo ischemia detection by luminescent nanothermometers. *Advanced Healthcare Materials* 6 (4): 1601195.
37. Du, P., Y. Hua, and J.S. Yu. 2018. Energy transfer from VO43—group to $Sm^{3+}$ ions in $Ba_3$ $(VO_4)_2$: $3x$$Sm^{3+}$ microparticles: A bifunctional platform for simultaneous optical thermometer and safety sign. *Chemical Engineering Journal* 352: 352–359.
38. Sales, T., et al. 2022. White light source and optical thermometry based on zinc-tellurite glass tri-doped with $Tm^{3+}/Er^{3+}/Sm^{3+}$. *Journal of Alloys and Compounds* 899: 163305.
39. Greenwood, G. 1956. The growth of dispersed precipitates in solutions. *Acta Metallurgica* 4 (3): 243–248.
40. Toy, R., et al. 2011. The effects of particle size, density and shape on margination of nanoparticles in microcirculation. *Nanotechnology* 22 (11): 115101.
41. Nanda, K., et al. 2015. Study of vibrational spectroscopy, linear and non-linear optical properties of $Sm^{3+}$ ions doped BaO–ZnO–$B_2O_3$ glasses. *Solid State Sciences* 45: 15–22.
42. Azlan, M., et al. 2018. Linear and nonlinear optical efficiency of novel neodymium nanoparticles doped tellurite glass for advanced laser glass. *Educational JSMT* 5 (2): 47–66.
43. Slimane, A.B., et al. 2013. On the phenomenon of large photoluminescence red shift in GaN nanoparticles. *Nanoscale Research Letters* 8 (1): 1–6.
44. Som, T., and B. Karmakar. 2009. Green and red fluorescence upconversion in neodymium-doped low phonon antimony glasses. *Journal of Alloys and Compounds* 476 (1–2): 383–389.
45. Wei, T., et al. 2015. Comprehensive evaluation of the structural, absorption, energy transfer, luminescent properties and near-infrared applications of the neodymium doped germanate glass. *Journal of Alloys and Compounds* 618: 95–101.
46. Dutchaneephet, J., A. Limpichaipanit, and A. Ngamjarurojana. 2019. Optical spectroscopic investigations of neodymium and erbium added bismuth silicate glasses. *Optik* 178: 111–116.
47. Azam, M., and V.K. Rai. 2018. Enhanced frequency upconversion in $Er^{3+}$–$Yb^{3+}$ codoped heavy metal oxides based tellurite glasses. *Methods and Applications in Fluorescence* 6 (2): 025002.
48. Azam, M., and V.K. Rai. 2017. $Ho^{3+}$–$Yb^{3+}$ codoped tellurite based glasses in visible lasers and optical devices: Judd-Ofelt analysis and frequency upconversion. *Solid State Sciences* 66: 7–15.
49. Azam, M., V.K. Rai, and P. Mishra. 2016. Enhanced frequency upconversion and non-colour tunability in $Er^{3+}$–$Yb^{3+}$ codoped $TeO_2$–$WO_3$–$Pb_3O_4$ glasses. *Journal of Materials Science: Materials in Electronics* 27: 12633–12641.
50. Azam, M., V.K. Rai, and D.K. Mohanty. 2017. Spectroscopy and enhanced frequency upconversion in $Nd^{3+}$–$Yb^{3+}$ codoped TPO glasses: Energy transfer and NIR to visible upconverter. *Methods and Applications in Fluorescence* 5 (3): 035005.

51. El-Mallawany, R. 1993. Longitudinal elastic constants of tellurite glasses. *Journal of Applied Physics* 73 (10): 4878–4880.
52. El-Mallawany, R., and G. Saunders. 1987. Elastic behaviour under pressure of the binary tellurite glasses $TeO_2$–$ZnCl_2$ and $TeO_2$–$WO_3$. *Journal of Materials Science Letters* 6 (4): 443–446.
53. Hussain, N.S., G. Hungerford, R. El-Mallawany, M.J.M. Gomes, M.A. Lopes, N. Ali, J.D. Santos, and S. Buddhudu. 2009. Absorption an demission analysis of $RE^{3+}$($Sm^{3+}$ and $Dy^{3+}$): Lithium boro tellurite glasses. *Journal of the Nanoscience and Nanotechnology* 9: 3672–3677.
54. El-Mallawany, R. 2000. Specific heat capacity of semiconducting glasses: Binary vanadium tellurite. *Physica Status Solidi A* 177 (2): 439–444.
55. El-Mallawany, R., L. Sharaf El-Deen, and M. Elkholy. 1996. Dielectric properties and polarizability of molybdenum tellurite glasses. Journal of Materials Science 31: 6339-6343.
56. Hager, I., R. El-Mallawany, and M. Poulain. 1999. Infrared and Raman spectra of new molybdenum and tungsten oxyfluoride glasses. *Journal of Materials Science* 34 (21): 5163–5168.
57. Peng, S., et al. 2015. Intense visible upconversion and energy transfer in $Ho^{3+}$/$Yb^{3+}$ codoped tellurite glasses for potential fiber laser. *Optical Fiber Technology* 22: 95–101.
58. Mahraz, Z.A.S., et al. 2013. Concentration dependent luminescence quenching of $Er^{3+}$-doped zinc boro-tellurite glass. *Journal of Luminescence* 144: 139–145.
59. Heinz, M., et al. 2018. Formation of bimetallic gold-silver nanoparticles in glass by UV laser irradiation. *Journal of Alloys and Compounds* 767: 1253–1263.
60. Azlan, M., et al. 2019. Upconversion properties of erbium nanoparticles doped tellurite glasses for high efficient laser glass. *Optics Communications* 448: 82–88.
61. Cheng, P., et al. 2017. Enhanced broadband near-infrared luminescence from $Pr^{3+}$-doped tellurite glass with silver nanoparticles. *Optical Materials* 73: 102–110.
62. Zuo, T., et al. 2010. The big red shift of photoluminescence of Mn dopants in strained CdS: A case study of Mn-doped MnS–CdS heteronanostructures. *Journal of the American Chemical Society* 132 (19): 6618–6619.
63. Ren, P., et al. 2014. Green photoluminescence from erbium-doped molybdenum trioxide. *Materials Letters* 122: 320–322.
64. Mahraz, Z.A.S., M. Sahar, and S. Ghoshal. 2015. Enhanced luminescence from silver nanoparticles integrated $Er^{3+}$-doped boro-tellurite glasses: Impact of annealing temperature. *Journal of Alloys and Compounds* 649: 1102–1109.
65. Raju, K.V., et al. 2013. Judd-Ofelt analysis and photoluminescence properties of $RE^{3+}$ (RE= Er and Nd): Cadmium lithium boro tellurite glasses. *Solid State Sciences* 15: 102–109.
66. Hamza, A.M., et al. 2019. Physical properties, ligand field and Judd-Ofelt intensity parameters of bio-silicate borotellurite glass system doped with erbium oxide. *Journal of Luminescence* 207: 497–506.
67. Elkhoshkhany, N., et al. 2018. Influence of $Sm_2O_3$ addition on Judd-Ofelt parameters, thermal and optical properties of the $TeO_2$–$Li_2O$–ZnO–$Nb_2O_5$ glass system. *Materials Characterization* 144: 274–286.
68. Slimane, A., et al. 2013. On the phenomenon of large photoluminescence red shift in GaN nanoparticles. *Nano Scale Research Letters* 8: 342.
69. Azlina, Y., et al. 2020. Optical performance of neodymium nanoparticles doped tellurite glasses. *Physica B: Condensed Matter* 577: 411784.
70. Duffy, J.A. 1989. Optical basicity of titanium(IV) oxide and zirconium(IV) oxide. *Journal of the American Ceramic Society* 72 (10): 2012–2013.
71. Sidek, H., et al. 2004. Formation and elastic behavior of lead-magnesium chlorophosphate glasses. *Turkish Journal of Physics* 28 (1): 65–72.
72. Dimitrov, V., and T. Komatsu. 2005. Classification of oxide glasses: A polarizability approach. *Journal of Solid State Chemistry* 178 (3): 831–846.
73. Sun, J., et al. 2011. Glass formation and properties of Ge–Te–$BiI_3$ far infrared transmitting chalcohalide glasses. *Spectrochimica Acta Part A: Molecular and Biomolecular Spectroscopy* 79 (5): 904–908.

74. Dimitrov, V., and S. Sakka. 1996. Linear and nonlinear optical properties of simple oxides. *II. Journal of Applied Physics* 79 (3): 1741–1745.
75. Hamza, A.M., et al. 2022. Borotellurite bio-silica glasses doped by erbium nanoparticles: Structural and thermal properties. *Journal of Theoretical and Applied Physics* 16 (2): 1–7.
76. Bouzidi, M., et al. 2022. Downconversion mechanism in $Er^{3+}/Yb^{3+}$ codoped fluorotellurite glasses to enhance the efficiency of c-Si PV cells. *Journal of Non-Crystalline Solids* 595: 121837.
77. Sailaja, P., et al. 2022. Near-infrared photoluminescence studies of neodymium ions doped $SrO–Al_2O_3–BaCl_2–B_2O_3–TeO_2$ glasses for laser and fiber amplifier applications. *Optics and Laser Technology* 156: 108569.
78. Morova, Y., et al. 2022. Tunable continuous-wave laser operation of $Tm^{3+}$ ion doped tellurite glass near 2 $\mu$m. *Journal of Luminescence* 252: 119318.
79. Wang, P., et al. 2022. Effects of melting temperature on color-changing in germano-tellurite niobate glass. *Optics and Laser Technology* 156: 108627.
80. Tang, D., et al. 2023. 2–3 $\mu$m mid-infrared luminescence of $Ho^{3+}/Yb^{3+}$ co-doped chloride-modified fluorotellurite glass. *Spectrochimica Acta Part A: Molecular and Biomolecular Spectroscopy* 285: 121833.
81. Li, C., et al. 2023. Broadband NIR radiative transitions in $Er^{3+}/Tm^{3+}$ co-doping tellurite glass material, Broadband NIR radiative transitions in $Er^{3+}/Tm^{3+}$ co-doping tellurite glass material. *Materials Research Bulletin* 158: 112042.
82. Patra, P., and K. Annapurna. 2022. Transparent tellurite glass-ceramics for photonics applications: A comprehensive review on crystalline phases and crystallization mechanisms. *Progress in Materials Science* 125: 100890.
83. Dolhen, M., et al. 2018. $Nd^{3+}$-doped transparent tellurite ceramics bulk lasers. *Scientific Reports* 8 (1): 4640.

# Index




R. El-Mallawany, *The Physics of Advanced Optical Materials: Tellurite Glasses*,
https://doi.org/10.1007/978-3-031-45245-1

The manufacturer's authorised representative in the EU is Springer Nature Customer Service Centre GmbH, Europaplatz 3, 69115 Heidelberg, Germany. If you have any concerns regarding our products, please contact ProductSafety@springernature.com

Printed and bound by CPI Group (UK) Ltd, Croydon, CR0 4YY
01/10/2026
02225880-0004